TORTOISES OF THE WORLD

GEORGE R. ZUG AND DEVIN A. REESE

TORTOISES

of the World Giants to Dwarfs

Johns Hopkins University Press
Baltimore

Printed in the United States of America on acid-free paper
9 8 7 6 5 4 3 2 1

Johns Hopkins University Press
2715 North Charles Street
Baltimore, Maryland 21218
www.press.jhu.edu

Library of Congress Cataloging-in-Publication Data
Names: Zug, George R., 1938– author. | Reese, Devin A., 1963– author.
Title: Tortoises of the world : giants to dwarfs / [George R. Zug and Devin A. Reese].
Description: Baltimore, Maryland : Johns Hopkins University Press, 2024. | Includes bibliographical references and index. |
Identifiers: LCCN 2023022277 | ISBN 9781421448350 (hardcover) | ISBN 9781421448367 (ebook)
Subjects: LCSH: Testudinidae.
Classification: LCC QL666.C584 Z84 2024 | DDC 597.92/4—dc23/eng/20230613

LC record available at https://lccn.loc.gov/2023022277

A catalog record for this book is available from the British Library.

Special discounts are available for bulk purchases of this book. For more information, please contact Special Sales at specialsales@jh.edu.

With gratitude to every hand and
mind that has contributed to
our knowledge of tortoise biology
and their survival in the wild

CONTENTS

PREFACE

What makes a giant and its opposite, a dwarf? Life comes in many shapes and sizes. In any plant or animal group with even modest diversity, there is a range of body sizes. Our human perception partitions this range into small, average, and large. No matter whether the organisms are ants or whales, we tend to sort their members by size and name the biggest ones accordingly: the Giant Squid, the Goliath Frog, the Giant Otter, or the Giant Huntsman Spider.

Eighteenth-century sailors visiting remote islands in search of fresh water and food encountered large land turtles and called them "giant" tortoises. For these seafarers, the tortoises were fascinating—albeit primarily as food. The sailors quickly discovered that the turtles could be kept alive for weeks, or even months, aboard ships. Long before refrigeration was invented, the animals' hardiness was wonderful for the sailors' provisions but devastating for the tortoises.

Still, some giant tortoise populations survived the hunting pressures of the 1700s to 1800s. Early naturalist explorers also discovered other species of tortoise, some small enough to fit in your palm. Since then, several centuries of recorded human interactions and research on tortoises have yielded a rich record of their behaviors, physiology, and other aspects of their life history. From the diminutive species to the gigantic, tortoises (family: Testudinidae) share a distinct suite of characteristics while also demonstrating an array of adaptations suitable to their specific environments.

We have long been enthralled by the scientific family of turtles known as tortoises. We invite readers to share our fascination by exploring the who, what, when, where, and why of these wondrous creatures.

We wrote *Tortoises of the World: Giants to Dwarfs* as a semi-technical book for a general audience of natural history readers. Our major goal was to present the full diversity of tortoise biology. To do that, we relied on the substantial body of scientific literature; hence our information sources were largely peer-reviewed articles. To serve general readers, avoid the disruption that occurs with in-text citations, and limit the size of the resulting book, we did not include a bibliography of our source literature. We recognize that some readers desire references. For those readers, we have created a source document and published it in the Smithsonian Herpetological Information Services (SHIS). This document, "Bibliography of *Tortoises of the World*," is available at the Smithsonian Library's website: https://repository.si.edu/handle/10088/842.

TORTOISES OF THE WORLD

CHAPTER 1

How the Tortoise Beat the Hare

Figure 1.1
The tortoise and the hare race.

All tortoises are turtles, but not all turtles are tortoises, much like all pineapples are fruits, but not all fruits are pineapples. Tortoises represent one of the 14 living families of turtles in the world (see Appendix Table A.1). The term "tortoise" is often used to refer to turtles that are not actually tortoises. Marine turtles used to be called tortoises, and in Australia, freshwater turtles have commonly been called tortoises, but these are both misnomers. For our purposes, "tortoise" refers exclusively to turtles in the taxonomic family Testudinidae, and all tortoises live strictly on land.

After the hare in Aesop's fable taunted the tortoise for his slowness, the tortoise proposed a race (Fig. 1.1). We're not certain which species of tortoise and hare were competing, but presumably they were Greek ones that lived during Aesop's time. Even though hares can run as fast as 50 km/hr compared to the plodding pace of a tortoise at 1–2 km/hr, the slow and steady tortoise improbably won the race. Biology offers some other wins for tortoises that are illuminated in this book.

Sneak Previews

Chapter 2. On the Body Plan

As the most terrestrial members of the order Testudines (turtles), tortoises have remarkable body plans. In contrast to hares and other mammals, a tortoise's shoulders and pelvic girdle lie inside its ribcage. Indeed, the ribs form a major part of the shells of all turtles. Their bony anatomy includes an external covering of keratin plates, called scutes, which are a novel adaptation common to all turtles. Tortoises stand out for their high-arched shells, which are fully domed in some species. Presumably, their shells are fine-tuned for resisting predators, although shape and especially height affect other aspects of their lives, including physiology and reproduction. Not only do the sturdy shells protect the animals' organs, but also most tortoises can fully withdraw their heads and necks within the shell and protect themselves with shieldlike forearms that are armored with thick scales. Their rears are protected by the thick soles of the hind feet that face outward when their hind legs are withdrawn. The arrangement of tortoise musculature, including extra-large leg muscles and tendons, reflects the challenges of walking on land while hauling around a weighty, inflexible shell. These high-efficiency muscles allow them to conserve energy, keeping their metabolic rates low even while moving around. How tortoises eat is also exceptional. In contrast to all other living reptiles, tortoises (indeed, all turtles) are toothless, munching on their plant-rich diets with sharp jaws made of the same material as bird beaks—keratin. In short, the tortoise body plan is an evolutionary innovation that's essential to their lifestyle in terrestrial habitats.

Chapter 3. On Resilience

Tortoises are extremely resilient, able to survive long periods without food or water in conditions that would quickly kill hares. Their resilience is grounded in unique adaptations of their internal body systems and physiology. A rigid shell requires a different breathing mechanism from that of any other terrestrial vertebrate because the rib cage can't expand and retract to draw air in and out of the lungs. So, all turtles use a bundle of posterior body muscles to inflate and deflate the pleural cavity like a bellows. They breathe infrequently, thanks to their higher threshold for carbon dioxide accumulation in the bloodstream. At their tolerated threshold, most birds and mammals would pass out, if not suffocate. Also, a tortoise heart does not pump steadily; instead, it pauses periodically as an individual goes about its slow routine of living. Tortoises—like camels—have adapted to limited access to water in many of their habitats. Tortoises store water in their large bladders for later use, extracting water as needed to keep bloodstream toxins secreted by physiological processes at safe levels. They digest food slowly, chomping with powerful jaw muscles to extract nutrients from plant materials. Like other reptiles, tortoises are ectotherms ("outside temperature"), getting most of their body heat from their environments. Tortoises shuttle between sunny spots on land to cooler places underneath shrubs or inside burrows. In short, tortoises are resilient because of their special systems to reduce oxygen use, conserve water, extract nutrients from plants, and behaviorally regulate their body temperatures.

Chapter 4. On Reproduction

Tortoises accomplish some unusual reproductive feats. Female tortoises, for example, can reproduce throughout their adult lifetimes, which may encompass a century. They can also store sperm for years, resulting in clutches of eggs from multiple fathers. In contrast, a hare is lucky to survive 10 years, and a female hare's eggs are fertilized soon after copulation. All tortoises (indeed, all turtles) excavate a nest cavity in which they lay eggs, as opposed to birthing live young like some lizards and snakes. Although many tortoise activities are quiet, courtship and mating are a noisy affair, with males grunting and roaring. In some species, males fight in a shoving match like bumper cars, and the losing male may get flipped over. Tortoises' aggressive courtship does not appear consensual on the female's end, but a female can resist mating by pressing the rear of her shell into the ground, which prevents a male from reaching his tail underneath to copulate. How many eggs a female tortoise lays in each nesting depends on her species, body size, and health. The sex of hatchling tortoises, at least in some species, depends on nest temperature during incubation, a phenomenon also seen in most other turtles, crocodilians, and some lizards. Typical of most reptiles, neither male nor female tortoises care for the hatchlings, although in a few species females have been observed guarding their nests. In short, tortoise reproduction is a vigorous process that includes male competition and some female choice in refusing copulation, selecting nest sites, and using stored sperm to fertilize eggs.

Chapter 5. On the Life Cycle

While a hare is born helpless, needing parental care, the first task for a turtle is to make its own way out of the egg. Tortoise eggs hatch after a handful of weeks to nearly a year, depending on the species. A hatchling slices open the leathery eggshell from the inside using a small, horny point on its beak. While gradually extricating itself from the egg and during its early days and weeks, the hatchling is sustained by the yolk sac. Many hatchling turtles navigate directly to waterways after emerging from the nest, but hatchling tortoises remain on land. As turtles grow, larger keratinous scutes develop beneath the smaller old scutes that cover the shell. Many turtles shed old scutes, but in tortoises they stack up like a pancake pyramid, resulting in treelike rings that provide a rough age estimate. Tortoises live a remarkably long time, with some giant tortoises reaching more than 175 years, a span of time that would encompass dozens of hare lifetimes. Tortoises are slow to reach reproductive maturity, but curiously, once they do, they continue to age slowly. In contrast to marine turtles that lay hundreds of eggs every year and for whom each egg has little chance of survival, tortoises lay a small number of eggs, a strategy that works if parents live a long time and offspring have better chances of survival. In short, tortoises have a life history strategy of slow growth with continual reproduction throughout most of their long lives.

Chapter 6. On Ecology

Tortoises overlap with hares in dry, grassland habitats, thus making their legendary race geographically plausible. Tortoises also occupy tropical forests. The broad temperature and humidity tolerances of the distinct species allow them to occupy habitats ranging from the equatorial Tropics north to central Asia and south to the tip of Africa. Although hares are famous for their burrowing, tortoises are also adept ecosystem engineers, digging shallow cavities under plants and in some species excavating deep burrows. Tortoise burrows provide a refuge from extreme temperatures and aridity, which also attracts a diverse array of other animals from insects to mammals to these refugia. A tortoise shell provides a formidable defense, but tortoises nevertheless provide food for predators that have sharp beaks and claws, including raptors, foxes, hyenas, and monitor lizards. Tortoise populations are structured by the distribution of refuges like burrows as well as the availability of food and shade resources. Their movement patterns are often driven by seasonal changes in resource availability, with some tortoises avoiding extreme conditions by hibernating or estivating during cold or dry periods. Tortoises mostly eat plant materials. As they graze through large quantities of fruits and leaves, tortoises maintain open habitats and disperse seeds in their scat. Despite their terrestrial habits and home ranges, tortoises sometimes move over long distances, including by floating across water bodies, which has resulted in their colonization of islands. In short, tortoises have a broad distribution on Earth and complex relationships with plants, animals, and other resources in their habitats.

Chapter 7. On Diversity

Hares (Leporidae) edge out tortoises on diversity, with 60 species of hare compared to about 47 living species of tortoise (family Testudinidae; see Appendix Table A.1). But tortoises inhabit all continents except Australia and Antarctica, whereas hares occur only on Africa, Eurasia, and North America. Modern tortoise diversity encompasses three lineages, with most living tortoises (all genera other than *Gopherus* and *Manouria*) in the Testudininae. Because tortoises have limited migration abilities (other than getting transported by humans or water), native species of tortoise tend to be geographically clustered. North America has several species of tortoises in the genus *Gopherus*, and South American tortoises are all *Chelonoidis*, including the giant Galápagos Tortoises living across multiple islands. A single genus—*Testudo*—accounts for all tortoises native to Europe, southwest Asia, and northern Africa. Tropical Asia harbors three genera of tortoises, and sub-Saharan Africa wins the diversity prize with nine genera, likely because it was a hub for the origin of modern tortoises, that is, for all tortoise species except the ancient *Gopherus* and *Manouria* lineages. The nearby Madagascar and other western Indian Ocean islands have three additional living species, remnants of the historically diverse tortoise fauna. Of the approximately 10 giant species in the Indian Ocean, only the Aldabra Giant Tortoise remains. Across these distinct regions of the world, tortoises vary in size, shape, life history, and behavior, though they maintain the same successful tortoise body plan. In short, tortoises are geographically clustered with a concentration of diversity in sub-Saharan Africa.

Chapter 8. On Origins

Tortoises were on Earth before hares, appearing at least 50 million years ago (mya). Hares appeared about 40 mya. Thanks to their hard body parts, turtles (including tortoises) have left a decent fossil record that tells the story of their origins. Turtles go back to at least 221 mya to species that lived on land during the Triassic, but they became mostly aquatic during the Jurassic. Later, tortoises evolved from aquatic turtles, with the first known tortoise-to-be venturing onto land during a Late Paleocene warm period (beginning 58 mya). A fossil of this ancestor species of all tortoises is yet to be discovered. The oldest tortoise fossils have been found in western North America, but molecular data place the origins of tortoises in Asia. From there, they spread to Europe and Africa, eventually colonizing the Americas. The earliest tortoises underwent evolutionary innovations in walking, breathing, and feeding that allowed them to thrive in terrestrial environments and become true tortoises. They altered their habitats as some of the first seed dispersers and diversified into a huge range of sizes, from the size of a person's palm to the whiskey-barrel-sized giant tortoises of the Pliocene and Pleistocene. Three tortoise lineages survived to the present: the Manouria lineage in Asia; the Gopherus lineage in North America; and the later Testudininae lineage, which includes most of the species living today. In short, tortoises evolved from swimming turtles that made their debut on land and spread around the globe while diversifying into a variety of shapes and sizes.

Chapter 9. On Decimation

Tortoises don't appear on dinner menus as often as hares, so you might conclude that they're more immune to the ravages of humans. In truth, tortoises have a long and hazardous history of interactions with people. For most of tortoises' long tenure on Earth, there were no humans. Tortoise populations grew and spread, colonizing continents and islands. The arrival of hominins several million years ago ushered in the most dramatic challenges tortoises ever faced. Tortoise diversity has declined from the prehuman tally of about 120 species to the current tally of fewer than 50. Human activities affect tortoises in a variety of direct and indirect ways. Ancient remains of tortoise bones show that humans were harvesting them for food as early as a million years ago. The giant tortoises inhabiting the Galápagos and western Indian Ocean islands were nearly hunted to extinction during the 18th–19th century once sailors, whalers, and other seafarers recognized their capacity to supply food and water aboard ships. Tortoises are still valued in some cultures for meat or perceived medicinal properties. During this past century, massive habitat changes, as well as a lucrative and unrelenting international pet trade, have devastated most tortoise populations. Introduced predators such as rats and dogs root out nests and hatchlings. Where tortoises remain, they face low genetic diversity from historic reductions in population size, as well as continuing obstacles to population growth. Their drawn-out maturation and reproduction make it difficult for tortoise populations to rebound even once obstacles are removed.

Chapter 10. On Conservation

Indigenous peoples' knowledge of—and in some groups, reverence for—tortoises, coupled with the general conservation movement that was launched in the 20th century, has offered protections to the remaining tortoise populations. The 1900 Lacey Act prohibiting commercial exploitation of wildlife, the establishment of national forests and other lands managed for hunting, and the passage of the Endangered Species Act were all instrumental to the future of tortoises. The initial protections for the large, charismatic Galápagos Tortoise were gradually extended to include a global suite of species. Because all tortoise species are imperiled, with many in danger of extinction, the Convention on International Trade in Endangered Species of Wild Fauna and Flora (CITES) offered them an international lifeline. Numerous research programs around the world have assembled the data that we draw on in this book, information that has also informed the development of tortoise conservation programs. Because the pressures on tortoise populations are complex and entangled with poverty, human conflict, resource limitations, and politics, conservation solutions are correspondingly multifaceted. Around the world, conservation programs have emerged that—for their success—rely on collaboration at multiple levels from communities to policymakers. Ongoing measures include establishing protected "tortoise villages," running captive breeding and reintroduction programs, starting awareness-building campaigns, and passing legislation to try to ward off the extinction of more species. In short, tortoises have taken a hard hit since humans began encroaching on their habitats; however, growing knowledge of their biology helps ensure their future on Earth.

The chapters of this book provide a deep look at tortoises based on the best biological knowledge to date about all aspects of their lives. Although Aesop's fable may have been poking fun at the slow walking speeds of tortoises, it also celebrates their persistence and resilience. Their myriad adaptations represent more than 50 million years of evolution. Tortoises stand out for their life history, physiology, and behavior specialized around the unique body plan of a hard, encasing shell that protects soft parts while limiting movement. Tortoises have passed the evolutionary test of time, but the very qualities that define them—slowness, hardiness, few defenses other than the shell—have also made them easy targets for human exploitation. Whether they'll pass the human-era test of time remains to be seen. The more we know about tortoises, the more we can all contribute to helping them pass their most challenging test yet.

CHAPTER 2

Life in a Shell

Tortoise Body Plan

It would be hard to mistake a tortoise for any other animal. Their chunky front legs and columnar hind legs, sturdy shells, and dry, scaly skin are giveaways, not to mention their lumbering gait, which is led by a beaked head on a long, wrinkly neck. This tortoise body plan, while divulging their membership in the broader group of turtles, is uniquely suited to these animals' special lifestyle. Natural selection has taken the body plan of their aquatic turtle ancestors and uniquely modified it for their special terrestrial lifestyle. The only other living turtles that have converged on some aspects of this land-adapted body plan are the more delicate box turtles (*Terrapene* spp.), the most terrestrial members of the family of pond turtles (Emydidae).

Uniqueness

Elephantine Legs

Tortoises (family Testudinidae) are turtles (grouped in the broader turtle order Testudines) and thus share many physical features with freshwater and sea turtles. But even a quick glance at a tortoise reveals a unique feature—hind legs that are column shaped and structurally similar to elephant legs. They end in broad, blunt feet, although it is difficult to tell where the legs end and the feet begin (Fig. 2.1). Their front legs are not elephantine, nor are they like the typical slender legs of other reptiles (including other turtles). Tortoises' front legs have a broad, muscular forearm distinctly set off from the narrower but still robust upper arm.

Neither tortoises nor elephants can move their toes independently because their toe bones are embedded in the tissue under the skin of their feet. From the outside, the only part of the toe visible is the nail—tortoises have five toenails on each front foot and three or four on each hind foot. On the bottom of the hind foot is a thick sole covered with cobblestone-like scales that are strengthened by keratin, the same protein that reinforces toenails. Tortoise hands and forearms are both also armored with an often-colorful layer of keratin scales.

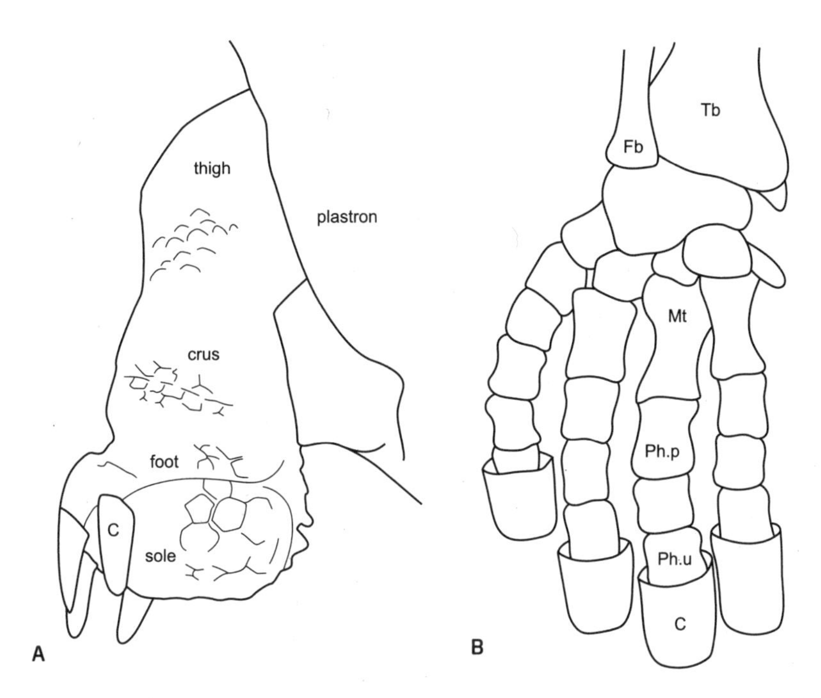

Figure 2.1
(A) Sketch of a Yellow-headed Tortoise (*Indotestudo elongata*) left hind leg from the rear and below, highlighting its elephantine appearance. **(B)** Hind foot skeleton of Aldabra Giant Tortoise (*Aldabrachelys gigantea*) viewed from the front. Abbreviations: C = claw; Fb = fibula; Mt = metatarsal; Ph.p = proximal phalanx; Ph.u = ultimate or terminal phalanx; Tb = tibia.
(A) US National Museum, USNM 222496; (B) Eléonore Dixon-Roche

Battening Down the Hatches

Aside from its head and tail, a tortoise's legs are its most exposed and vulnerable parts. When a tortoise withdraws into its shell, it pulls in its front and hind legs. The thick soles of its feet face outward and become the only exposed part of the hind legs, forming a barricade over the softer tissues. The elbows of its front legs nearly touch in front of its head, so that the thickly scaled skin shields the front opening. When facing a predator, a tortoise can batten down the hatches to protect itself (Fig. 2.2). Still, many predators manage to feed on tortoises, especially on hatchlings (see Chap. 6).

Heads Built for Chomping

Tortoise skulls are distinct in ways that relate to their generally herbivorous diets. Turtles, broadly, have solid skulls with no openings (fenestrae) other than for their eyes and nostrils. The top of the tortoise skull, however, is especially cut back (emargi-

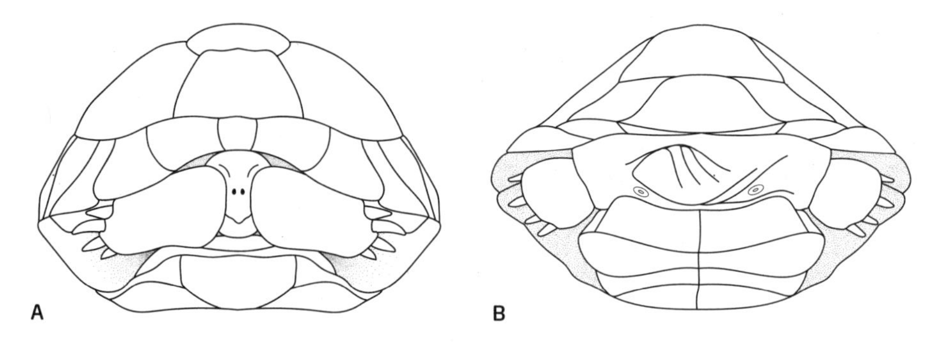

Figure 2.2
(A) Front view of a Hermann's Tortoise (*Testudo hermanni*) fully withdrawn into its shell.
(B) Rear view of a Spur-thighed Tortoise (*T. graeca*) fully withdrawn. Schematic outlines from photographs of US National Museum specimens; concept from Schleich and Kästle (2002)

nate), likely to accommodate the extra-large jaw muscles the creatures need for chewing up plants. Also, tortoises have a high, arched palate (roof of the mouth), which provides plenty of vertical room in their mouths for wads of vegetation (Fig. 2.3).

Modern turtles have no teeth at all, rather beaks. Tortoises have particularly sharp-edged sheaths on their upper and lower beaks. These keratinous sheaths, known as tomia, can cut through plant materials like a knife thanks to their serrated edges. Although the tomia do not have touch receptors, under each lies a layer of tissue with bundles of sensory nerves that receive signals from the jaw. Their jaw muscles are big, but tortoise neck muscles are modest compared to other, carnivorous turtles that use their jaws for prey capture or defense.

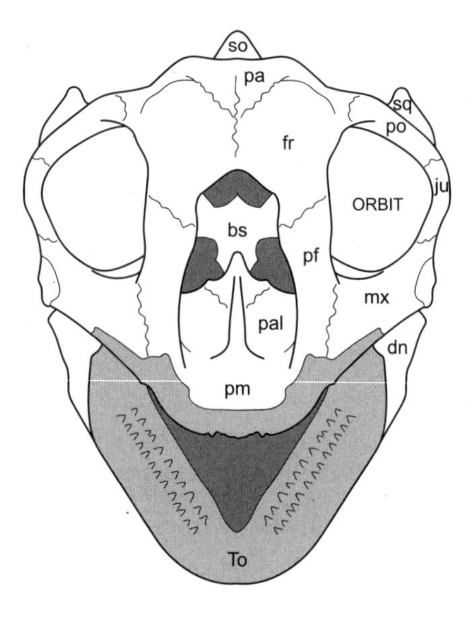

Figure 2.3
Front and slightly above view of skull and jaw of an Aldabra Giant Tortoise (*Aldabrachelys gigantea*). Bone is unshaded; keratinous sheaths (tomia) of upper and lower jaws are shaded; and open spaces within the skull are black. Abbreviations: bs = basisphenoid; dn = dentary; fr = frontal; ju = jugal; mx = maxillary; na = nasal; pa = parietal: pal = palatine; pf = prefrontal; pm = premaxillary; po = postorbital; so = supraoccipital; sq = squamosal; To = tomium. US National Museum, USNM 222496

Turtle Shells Generally

In evolutionary terms, turtles have a highly "conserved" body plan that has not changed dramatically over geologic time. Despite a turtle shell being one of the strangest vertebrate features, it has persisted for more than 240 million years, albeit with structural modifications in different turtle lineages. Many body functions that work in conjunction with the shell—such as ways of breathing, moving around, and reproducing—have a long shared evolutionary history.

Turtle shells are strong, an adaptation to protect against the extreme mechanical forces of predator claws and teeth. Turtles also face blunt force impacts, such as when bird predators drop them from high places onto rocks to fracture their shells. Turtle shells are innovative, impact-resistant structures composed of layers of specialized materials.

Bony Building Blocks

All turtles, whether terrestrial or aquatic, have an upper shell (carapace) and lower shell (plastron). No matter the type of turtle, the components and organization of the shell are fundamentally the same. The entire shell consists of about 50 bones connected to one another via a bony bridge (strut) on each side that keeps the shell from collapsing.

A turtle's plastron protects its underside with a set of nine bony plates: four left and right pairs, and a small median plate at the front end. On each side, the two middle pairs of bony plates curve upward to form the bridge to the edge of the carapace, uniting the bottom and top shells into a single, firm structural unit (Fig. 2.4).

The typical turtle carapace has a construction of three sets of bony plate building blocks: a midline series that goes over the top (one nuchal at the front followed usually by eight neurals and a final pygal bone at the tail end), eight bony plates that cover the sides (costals), and a ring of flat bones forming the outer edge (peripherals). All the carapace bones are linked like articulated tiles, but not fused, because their outer edges grow as the turtle grows. The ring of interlocked peripherals helps hold all the bones together in an array like puzzle pieces (Fig. 2.4).

Looking inside the shell of a turtle (minus the soft tissues), you'll see that the vertebrae of the trunk are attached to the bones of the carapace. Literally, the bones that enclose the spinal cord (neural arches) are fused to the neural plates. From the midline, the ribs extend outward and effectively disappear into the surface of the costal plates. As such, the vertebrae and ribs form a supportive framework for the carapace bones, like the frame of a building for its walls and roof.

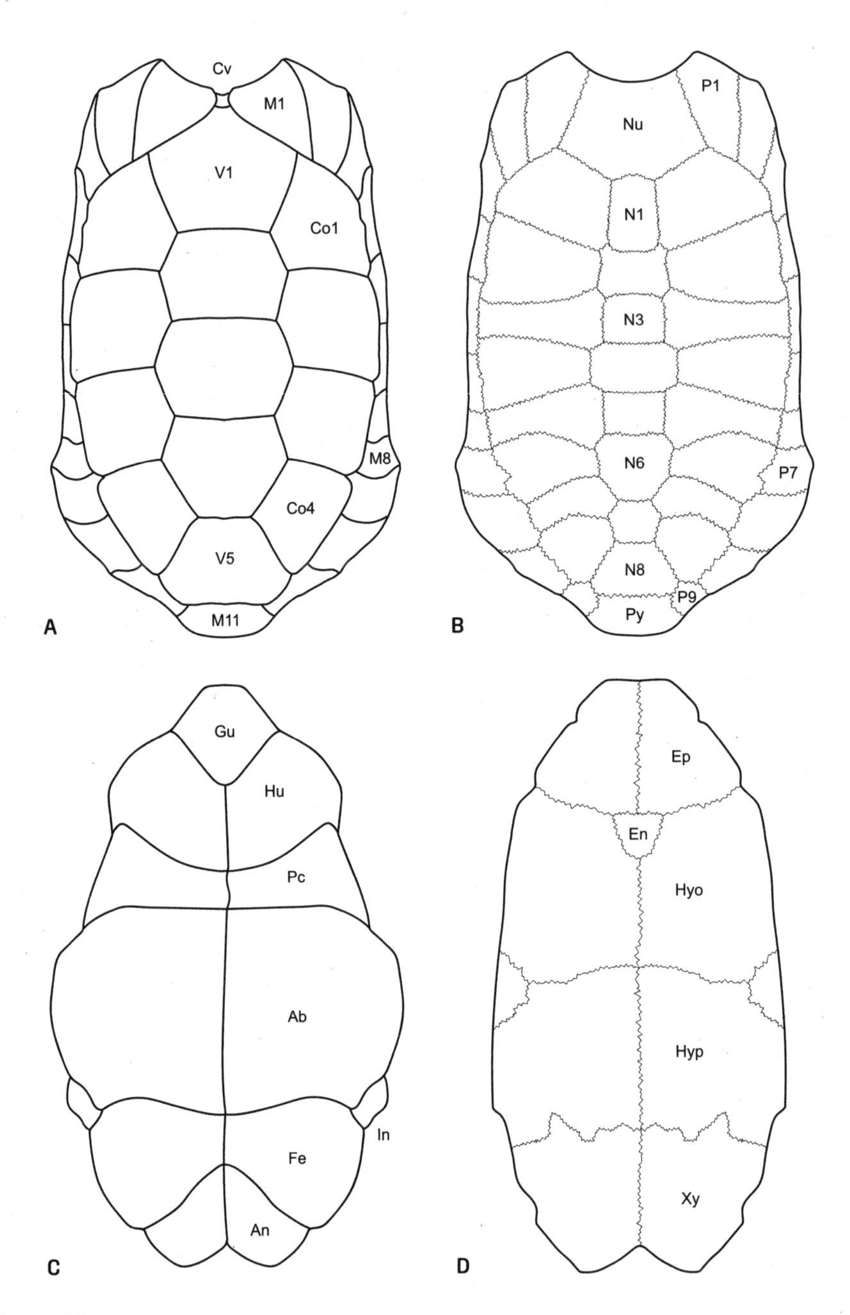

Figure 2.4
Shell of an Angulate Tortoise (*Chersina angulata*). **(A)** Carapace with scutes (Abbreviations: Co = costals; Cv = cervical; M = marginal; V = vertebral). **(B)** Carapace without scutes (Abbreviations: N = neural; Nu = nuchal; P = peripheral; Py = pygal). **(C)** Plastron with scutes (Abbreviations: Ab = abdominal; An = anal; Fe = femoral; Gu = gular; Hu = humeral; In = inguinal; Pc = pectoral). **(D)** Plastron without scutes (Abbreviations: En = endoplastron; Ep = epiplastron; Hyo = hyoplastron; Hyp = hypoplastron; Xy = xiphiplastron). (A), (B), and (C) after Loveridge and Williams (1957, Fig. 37A and B); (D) US National Museum, USNM 63018

Strengthening Architecture

A turtle shell is further protected by a covering of hard scutes that, like its scales, are made of keratin. The scutes—typically 38 on the carapace and 16 on the plastron—adjoin one another without overlapping. The scute boundaries are offset from the boundaries of the bones, much like how layers of bricks are offset for a wall, a much stronger architecture than if they were aligned. Beneath the dead scute layer is a durable living layer (the epidermis and dermis) that keeps the scutes securely connected to the bones. Turtle scutes are an evolutionary novelty not seen in any other organisms, yet they're present even in the earliest known turtles (Chap. 8).

A feature that has likely allowed turtles to persist for millions of years is the adaptability of their shells. Turtle shells have multiple adaptations, including size differences and variations in shell thickness (from thin and flexible to thick and rigid). As one of the most distinct vertebrate features, the turtle shell has persisted for more than 240 million years, albeit with structural modifications in different turtle lineages. In evolutionary terms, we say that turtles have a highly "conserved" body plan. Many body functions that work in conjunction with the shell, such as ways of breathing and reproducing, have a long shared evolutionary history.

Tortoise Shells Specifically

As they grow, many turtles shed scutes and replace them with new, larger ones formed as dermal cells underneath die and keratinize. Tortoises, however, retain their scutes, which therefore stack up into pyramids on the carapace over time. If you look down from the top, you see the entire small scute on the top, with concentric rings from other scute edges peeking out below it. A tortoise's plastron does not shed scutes either—they accumulate into a stack—but because they get worn down from friction against the ground, the multiple scute layers are not as evident (Fig. 2.5).

Counting the Rings

If you look down at one of the scute sections of a tortoise carapace, you'll see the exposed outer edge of each new scute as a ring (annulus) around the scute above it. You can count these annuli like tree rings to estimate a tortoise's age. In general, one scute layer is formed per year, although any growth spurt may create a new scute. No new scute material grows during periods of little or no growth, such as winters in a temperate climate. Thus counting scute annuli provides just a ballpark age estimate. As a tortoise matures, growth slows, causing the annuli to become narrower. Eventually, the incremental growth is so slight that counting the annuli becomes nearly

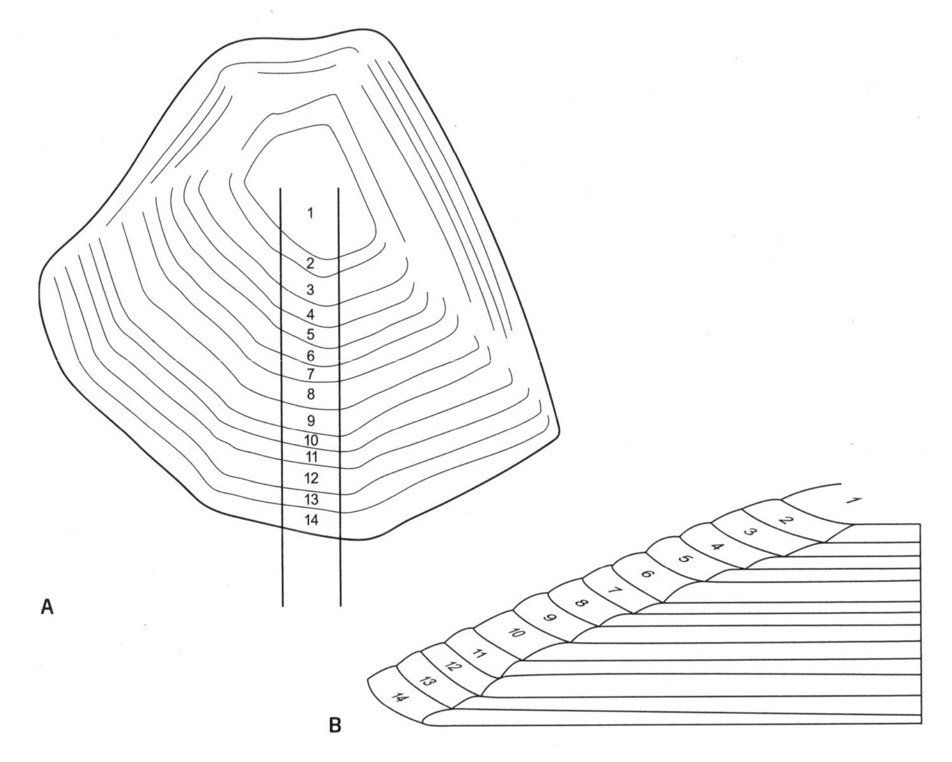

Figure 2.5
Growth of carapacial scutes displaying the successive stacking, typically annual in wild tortoises. **(A)** Left first costal scute of a Tent Tortoise (*Psammobates tentorius*) showing a pyramid of scutes topped by the scute it bore as a hatchling. **(B)** Schematic cross-section of the preceding costal plate shows the persistence of scutes from hatchling to death at approximately 15 years. (A) US National Museum, USNM 39418; (B) C.M. and G.R.Z.

impossible. In an old turtle, some annuli may be gone as the top, older scutes have worn away.

Natural Waterproofing

Tortoise scutes vary in thickness depending on the rate and duration of tortoise growth periods, which depend on climate and nutrition. Overall, because tortoise scutes are not shed and therefore keep growing, they're thicker than the scutes of other turtles. These thick, keratinous scutes are a waterproofing barrier for the tortoise, to some degree keeping water out but more importantly preventing the loss of body water by evaporation. Thanks to this barrier, many tortoise species can live in dry environments with little or no water available for drinking (Chap. 7). Tortoises have also evolved physiological mechanisms for conserving water, but they still lose water from their lungs via respiration.

Bio-Shield

The layer of interlocking hard and soft materials on a turtle—especially thick on a tortoise—form what engineers call a mechanical bio-shield, which provides excellent protection against a predator's sharp teeth or a long fall from a raptor's talons. Besides mechanical defense, tortoise shells serve many vital roles from camouflage, buoyancy, and thermoregulation to the regulation of metabolites such as fat, calcium, and ions.

A tortoise's shell also comes with liabilities relative to the more standard vertebrate body plan, however. In other vertebrates, including humans, the shoulder and hip bones are outside the expandable rib cage. In turtles—including tortoises—the ribs are affixed to the rigid shell, with the shoulder bones (pectoral girdle of clavicle and scapula) and hip bones (pelvic girdle of ilium, pubis, ischium) inside the rib cage, a novel architecture that required adaptations for moving and breathing (Chap. 3).

Shell Shape and Size

During the voyage of the *Beagle*, Darwin visited the Galápagos Islands in 1835. Invited to dine with the acting governor of these islands, he asked about an array of tortoise shells displayed on the governor's walls. Darwin reported in his book about his travels that the governor "can on seeing a tortoise pronounce with certainty from which island it has been brought." Whether or not the governor's bragging was accurate, Darwin soon discovered for himself that the islands harbored tortoise populations with different shell shapes ranging from distinctly domed to more saddle-backed.

Shell Shape

The terms domed, saddle-backed, and intermediate, first used to describe the Galápagos tortoises (*Chelonoidis niger* spp.), also apply to other tortoise species. These descriptors are based on the appearance of the carapace viewed from the side, called the lateral profile. The domed shape is semicircular, bounded on the bottom by the flat plastron. The saddle-backed shape is a stretched-out dome, with the front quarter of the carapace curving upward like a classic Spanish saddle. Intermediate domes include various other elongated carapace shapes, which tend to be rectangular in profile (Fig. 2.6).

The common description of tortoise shells as being "domed" is overused, given the diversity of shell shapes. In the absence of a consistent and accurate description

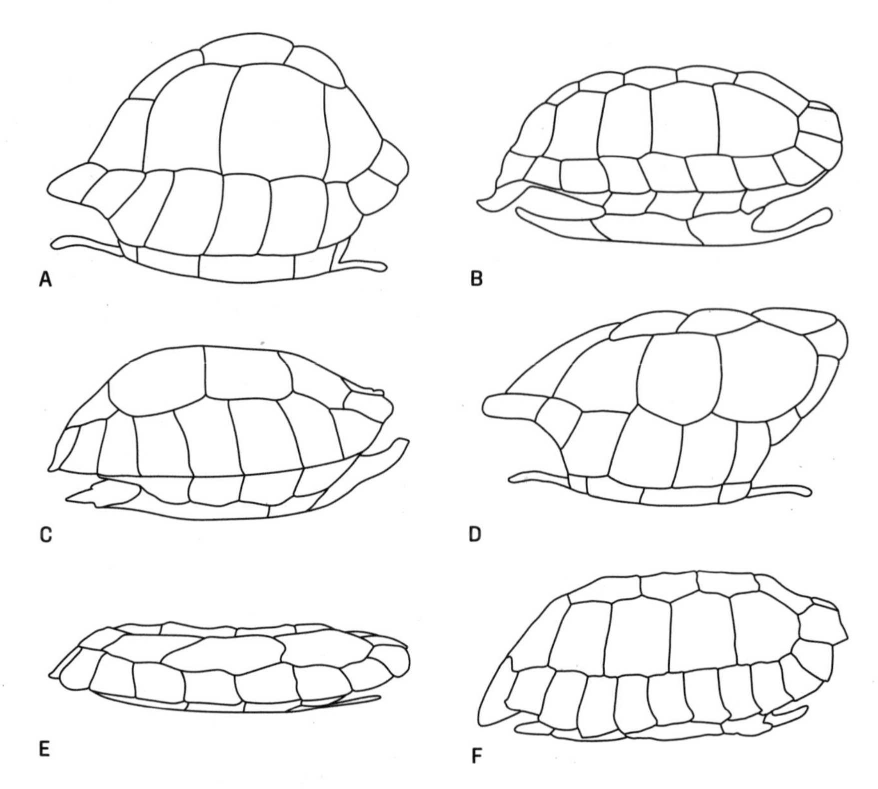

Figure 2.6
Outlines (side profiles) of the major shell shapes of the world's tortoises. **(A)** Domed (*Chelonoidis n. vandenburghi*). **(B)** Rectangular-flat (*Gopherus agassizii*). **(C)** Rectangular-domed (*Homopus areolatus*). **(D)** Saddle-backed (*C. n. hoodensis*). **(E)** Flat and broad (*Malacochersus tornieri*). **(F)** Rectangular-elongate (*Indotestudo elongata*). These shapes are not to scale but are proportionately correct. (A) and (D) after Zug (2013, Figs. 29A and C); (B), (C), and (E) from photographs of US National Museum specimens; (F) after Zug (2022, Fig. 35E)

of tortoise shapes, we developed a terminology based on the relative dimensions of the height and width of the carapace in relation to its total straight-line length. We have not incorporated the dimensions of the plastron because it's largely flat and does not change the shape of a tortoise in profile, although the bridge (which largely consists of plastral bones) does affect the relative height of the shell.

Here we define domed as a height that is at least 50% of its length and a width at least 60% of its length (Table 2.1). Domed carapaces occur in just over one-third of tortoise species. The saddleback shape applies to several Galápagos subspecies but not to any other living tortoise species, although a variation of the saddlelike shape develops occasionally in other tortoises. The remaining rectangular shell shapes pertain to most living tortoises. They vary in length and have flat to slightly arched tops of the carapace. Nevertheless, all living tortoise species except the flattened Pancake Tortoise

(*Malacochersus tornieri*) and the gently domed South African Dwarf Tortoises (*Chersobius* spp., *Homopus* spp.) have high shells with large internal volume.

Looking from the side does not reveal everything about carapace shape because some species are wider than others, such as the Bolson Tortoise (*Gopherus flavomarginata*), or the Pancake Tortoise compared to the Sulawesi Tortoise (*Indotestudo forestenii*) (Table 2.1). Also, even though a tortoise species has a predominant shape, variable carapace shapes occur in most species, particularly when raised in captivity. The variability within a natural-occurring species is likely dietary, such as the longer and narrower-shelled Aldabra Tortoises that are migratory and spend part of each

Table 2.1.
Height and Width Proportions of the Five Generalized Carapace Shapes of Living Tortoises

Species	CH/SCL (%)	CW/SCL (%)	CH/CW (%)	SCL (cm)
Domed				
Aldabrachelys gigantea	55–57	60–68	85–92	59–114
Astrochelys radiata	54–58	76–77	70–76	24–40
Chelonoidis n. abingtoni	45–52	63–74	70–76	44–92
Chelonoidis n. porter	54–58	76–77	68–78	65–151
Testudo graeca	53–55	71–78	68–76	8–34
Rectangular-flat				
Gopherus agassizii	41–46	75–80	55–57	12–37
Kinixys erosa	40–44	62–68	64–66	18–40
Rectangular-domed				
Gopherus berlandieri	47–53	76–77	61–68	13–24
Homopus areolatus	35	73	48	10–13
Kinixy belliana	40–46	62–71	56–73	14–22
Rectangular-elongate				
Indotestudo elongata	41–49	57–67	61–85	18–39
Indotestudo forestenii	37–40	65–69	57–58	11–27
Flat and broad				
Malacochersus tornerii	21–34	76–81	28–29	13–18

Note: Proportions are presented to demonstrate the differences that define the different shapes of tortoises, excluding the saddleback shape. Proportions are given as percentages, and SCL is the range of adult females and males in centimeters. Abbreviations are as follows: CH, carapace height at the middle of the second vertebral scute; CW, carapace width at the same transect; SCL, longest straight-line distance from the front to back edge of the shell parallel to the midline. A higher CH/SCL indicates a shell that is tall relative to its length. A higher CW/SCL indicates a shell that is wide relative to its length. And a higher CH/CW indicates a shell that is high relative to its width.

year in the scrub forest versus the more sedentary tortoises living on the grass plains. Although this terminology is not entirely satisfactory, it provides a way to describe and depict the diversity of tortoise shell shapes.

The shape of the marginal scutes (ringing the carapace edge) and the size of the anterior (front) and posterior (back) shell openings also dictate tortoise shell shape, commonly characteristic to a species or species group. For example, the huge anterior opening of the Galápagos saddleback carapace is accentuated by the upward flare of marginals around it, which perhaps allows space for the neck to extend upward.

Are the necks of saddlebacks extralong, or does the way they extend their necks fully to feed make them appear longer? An early study of Galápagos tortoise anatomy demonstrated that saddlebacks had significantly longer necks. More recently, researchers hypothesized that longer necks evolved as an adaptation for righting themselves when overturned. Saddleback Galápagos Tortoises live on volcanic islands with rocky, irregular surfaces and may be more likely to lose their balance. Although a morphometric and simulation study refuted the righting hypothesis, the long necks of saddleback tortoises are clearly advantageous for browsing on vegetation above their heads, and males stretch out their necks during dominance interactions (Chap. 4). Multiple uses for a characteristic may enhance a tortoise's survival advantage.

Saddleback shapes are not seen on mainland continents where large predators occur. And Galápagos saddleback tortoises are consistently smaller than their domed cousins, perhaps because of the harsher climates and more limited resources on the islands they inhabit. Most continental species of tortoises have modest openings in front (anterior) that their broad forelegs, protected by hard and thick scales, can fully block (Fig. 2.2). The rear openings range from modest to minimal, large enough to allow the hind legs to move but imposing constraints on the size of eggs females can lay (Chap. 4).

Shell Size

The size of a tortoise's shell, especially the carapace, is relevant for withstanding predator attacks. The smaller an object, the easier it is for a predator to get its jaws around it and apply enough multiple-point compression to break the shell. The minimum size (diameter and length) to thwart a predator's bite will depend on the species of carnivores encountered by a tortoise. For example, mountain lions (*Puma concolor*) can consume Berlandier's Tortoises (*Gopherus berlandieri*), and there are several records of predation by jaguars on South American tortoises (*Chelonoidis* spp.). But many tortoises are above the minimum length—about 20 cm (the length of a dinner fork)—to resist predation.

The significance of carapace size in defense is reinforced by the fact that predators target hatchlings and young turtles whose shells are too small to prevent predation. Once their carapace length reaches about 20 cm, they gain protection from most predators. A Central American Indigo Snake (*Drymarchon melanurus*) that tried to swallow a 47-mm-long hatchling Sinoloan Thornscrub Tortoise (*Gopherus evgoodei*) gave up after repeated attempts because it was just too big and rigid. Most adult tortoises are longer than 20 cm. Adults from very small species, however, such as the dwarf tortoises of southern Africa, are also common prey of raptors, crows, and even ostriches. A South African farmer observed that a single pair of Pied Crows (*Corvus albus*) fed 160 small Geometric Tortoises (*Psammobates geometricus*) to their nest of young.

Among living tortoises, the size range is staggering. Measuring the straight horizontal line (straight carapace length, or SCL) from the front edge of the carapace to the back edge, some species have adult SCLs of just 10 cm (the size of a playing card), whereas a few species exceed 1 m (the length of a baseball bat). The small ones are labeled "dwarfs" and the large ones "giants," but at what size do tortoises merit each of those labels? Because most species of tortoises have a maximum SCL of 21–40 cm (Fig. 2.7), here we categorize tortoises less than 20 cm as small and tortoises greater than 60 cm as large. The Karoo Dwarf Tortoise (*Chersobius boulengeri*) has a maxi-

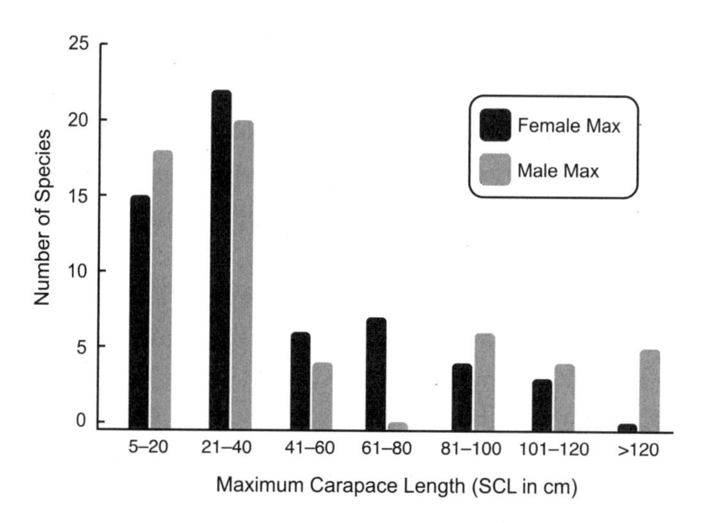

Figure 2.7
Bar graph of the maximum straight carapace lengths (SCLs, in centimeters) of the living species of the world's tortoises. The different populations or subspecies of the Galápagos Tortoises are included as individual entries. Female carapace lengths are shown as dark bars on the left of each paired column; male lengths are shown as gray bars on the right. The bar diagram is based on the maximum size data in the species accounts in chapter 7

mum adult SCL of 16 cm for females and 13 cm for males. A benchmark for "giants" is the maximum size of the smallest subspecies of Galápagos Tortoise, *C. n. guntheri* (SCL of 74 cm). Although these size cutoffs are somewhat arbitrary, defining them is important for talking about tortoises.

Moving with a Shell

Tortoise movements include walking, arguably running, and digging. They also include a variety of movements associated with courtship and mating. The leg bones of tortoises tend to be more robust than those of aquatic turtles, which makes sense because on land their bones must counter the force of gravity to hold the shell above the ground. The twisting forces that operate during movements could also place significant loads on their legs. Tortoises have cylindrical bones, an optimal shape for reducing twisting (torsion), versus the somewhat more flattened bones of aquatic turtles, which can adjust to neutral buoyancy to eliminate weight on their legs.

Walking Gait

Tortoise locomotion consists of sequential leg movements that carry the animal from one place to another. This may be as short as a single step between plants while grazing or as long as several kilometers when moving to a more productive feeding area or searching for a mate. The length of any turtle's step is limited by the size of the shell opening. The rearward movement of the front legs is restricted by the front edge of the bridge, and the forward movement of the hind legs by the back edge of the bridge.

Nevertheless, the turtle walking gait follows the same principles as any other four-legged (tetrapod) animal, principles that maintain balance in the most efficient manner when an animal's body is off the ground. The basic tetrapod walking gaits are the diagonal sequence gait and the lateral sequence gait. Both gaits have diagonal support; that is, a front foot and its diagonally opposite (contralateral) hind foot are on the ground at the same time.

Tortoises (and all other turtles) use the lateral sequence gait, in which both legs on the same side move and then the legs on the other side. For example, the right hind foot touches down—right front leg steps forward—right front foot touches down—left hind leg steps forward—left hind foot touches down—left front leg steps forward—left front foot touches down—right hind leg steps forward, and so on (Fig. 2.8). Most of the time, three legs are in contact with the ground, making a tripod that stabilizes the tortoise against pitching from side to side (Plate 2).

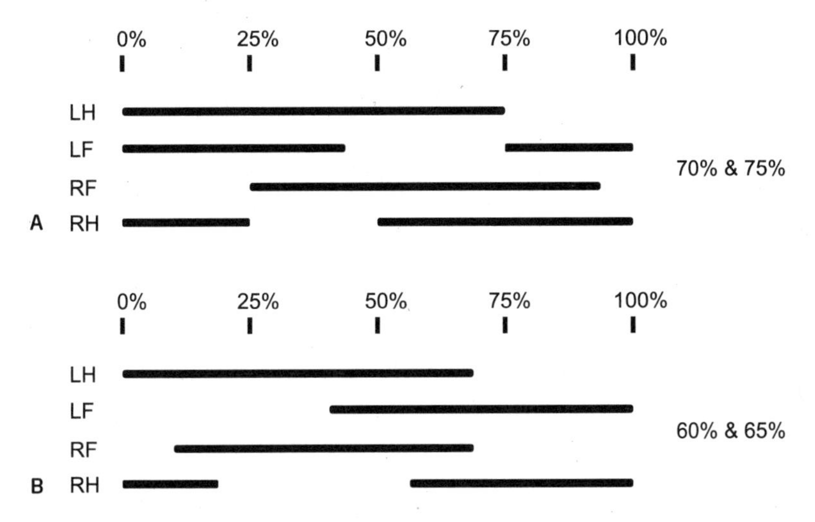

Figure 2.8
Gait diagrams of a single stride for a tortoise walking at (A) moderate speed and (B) fast speed. At a moderate speed, a tortoise maintains a tripod of support throughout a stride, whereas at a faster gait, there are periods of instability with only bipedal support. The dark bar indicates that the foot is on the ground; the blank space indicates a foot lifted and moving forward, that is, a step. For most tortoise gaits, no matter the speed, the hind foot remains on the ground slightly longer than the forefoot. These hypothetical diagrams show the forefoot on ground for 80% and hind foot for 86% of stride during a moderate walking speed, and 70% and 76%, respectively, for a fast walk. Abbreviations: LF = left forefoot; LH = left hind foot; RF = right forefoot; RH = right hind foot. Gait diagrams constructed by G.R.Z.

Generally, a tortoise increases its walking speed not by taking bigger steps (because it can't) but by taking faster steps. Male tortoises can cover more ground than females, even at a slower gait. Therefore, although a male tortoise may walk more slowly than a female, he gains distance from the increased length of his steps and stride, which is a matter of body mechanics.

Walking Mechanics

Although the sequence of leg movements and foot placements of tortoises matches that of walking and slow running gaits of lizards and mammals, the mechanics differ. Other tetrapods not constrained by shells bend their torsos as they walk, and especially as they run. Picture a lizard scurrying up a tree trunk. Its torso flexes laterally (from side to side), which lengthens its steps as well as pulls its body forward as the curvature of the body changes direction. Mammals that walk on four legs flex their torsos vertically (up and down) to gain step length and forward movement. Picture how the spine of a cheetah changes as it runs.

In contrast, a turtle's body cannot flex as it locomotes. Thus its walking movements are entirely dependent on muscles attached to the legs, most of which are connected to the shoulder bones (pectoral girdle) and hip bones (pelvic girdle). Further, a turtle has fewer trunk muscles than either a lizard or a mammal. Some trunk muscles were lost during the evolution of the turtle shell, whereas others have evolved to serve different functions. The four abdominal support muscles that remain—the transverse and oblique muscles on each side—are the major muscles used in turtle breathing (Chap. 3).

Shoulder and Hip Modifications

With leg movement limited by the rigid shell, the shape and structure of shoulder and hip bones had to evolve as well. Other reptiles have three shoulder bones—the clavicle, scapula, and coracoid—whereas the turtle shoulder girdle is simplified to two bones (Fig. 2.9), with the scapula taking on a new role of connecting to the shell to provide support. A lengthened rod part (acromion process) of the scapula extends horizontally medially, and a paddlelike bone (coracoid) extends diagonally rearward. Broad muscles anchor the acromion and coracoid to the plastron. The remaining shoulder girdle muscles attach to the upper arm bone (humerus) to move the front legs. The left and right sides of the shoulder girdle are independent, each pivoting forward and sideways during walking.

Each front leg moves in a roughly 90° arc from beside the neck to the bony bridge, with the upper arm staying nearly horizontal and the forearm and forefoot nearly vertical to hold the tortoise's body off the ground. The left and right sides of

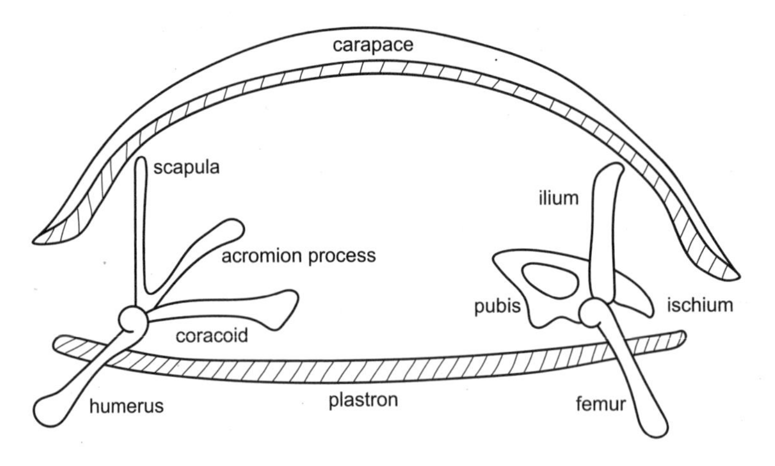

Figure 2.9
Sketch of the left limb girdles and proximal limb skeletons of a tortoise in a cutaway of the shell from a side-back aspect. Sketch by G.R.Z.

hip bones (pelvic girdle) are connected as in a mammal; the paired ischia and pubic bones form a flat, stable base. Muscles and ligaments anchor the base to the bottom shell (plastron), and the larger, top bone of each hip (ilium) extends upward to anchor to the underside of the carapace and ribs.

With the girdles inside the shell and leg movement constricted by the shell bridge, the upper arm bone (humerus) and thigh bone (femur) cannot swing in all directions as they otherwise would. The humerus swings forward and to the side, and the femur swings backward and to the side. Both the humerus and femur curve upward so they don't hit the plastron and give space for the lower leg bones to move without hitting the plastron. Basically, turtle shoulders, hips, and legs are substantially altered to making walking with a shell possible.

Digging

Tortoises dig! They dig with their front legs to make shallow depressions (pallets) and deeper burrows, and they dig with their hind legs to excavate egg chambers. Thanks to their strong, muscular forearms and sturdy claws, digging a shallow resting palette is feasible even in hard or gravelly soils. Rather than holding the front legs vertically as when walking, a tortoise digging a pallet holds its front legs horizontally and starts each backward-sweeping stroke from beside the head (Plate 3). The front legs alternate from side to side, while the hind legs typically push the plastron forward, resulting in a bulldozer effect. Once the depression is deep enough for the tortoise to sense a difference in soil moisture or temperature, the digging stops. A resting pallet has been formed.

Only a few species dig burrows (tunnels) that are sufficiently deep and long to create a stable microclimate (Chap. 6). A few species of tortoises don't dig at all, for example, if habitats are too rocky. The non-digging species tend to have more flexible, slender front legs with pointed, cone-shaped claws.

Nearly all turtles—including tortoises—dig their egg chambers with their hind legs despite their elephantine feet that lack flexible toes. The digging is laborious but nonetheless has worked successfully for about 50 million years.

Flipping Over

Tortoises may get rolled on their backs during mating, aggressive social interactions, predator attacks, or simply by losing their balance while traversing irregular terrain (Plate 3). The rigid shell of a tortoise handicaps it when it gets rolled over onto its back. Yet being able to get back on their feet is essential for tortoises to breathe effectively and avoid sun exposure. Despite their relatively short legs, tortoises right themselves by rolling over on the long axis of the shell, although it is energetically costly. A study of Spur-thighed Tortoises (*Testudo graeca*) found that the energy used for self-righting is double what's needed for walking.

Behavioral experiments and mathematical modeling have shown that a domed shell is an optimal shape for a tortoise to right itself. Even though not all tortoises have the "classic" domed shell (Table 2.1), all tortoises can right themselves. A long carapace also assists in righting by shifting the center of gravity closer to the plastron, increasing the chance that swinging legs will cause the body to roll back over. Often in male-male interactions, the upright male continues to aggressively ram the upside-down male and inadvertently assists its competitor in righting itself. It has been proposed that the larger, flared scutes at the tail end of male tortoises may make it harder to self-right but confer stability for mating.

CHAPTER 3

Physiology and Behavior

Tortoise Resilience

Slow and steady, in movement and in all aspects of their lives, must be the motto of tortoises. If a tortoise can avoid the hazards it may encounter as a hatchling or juvenile and reach adulthood, it has a high probability of surviving for at least several decades and often longer. Tortoises are famous for their longevity, which has contributed to their reputation as sturdy and resilient. Behind their resilience is a special physiology and set of behaviors associated with having a shell.

Cartoons often depict the tortoise shell as a protective house from which a naked tortoise can come and go. Nothing is further from the truth! The shell is an integral and permanent part of a tortoise, and it is the critical structure for their unique suite of time-tested physiological adaptations. As noted in Chapter 2, the shell underpins their mode of locomotion. Other unique tortoise behaviors relate to their needs for life on land, such as digging extensive burrows or wedging into rock crevices. Behaviors reflecting social hierarchies are common to all turtles but are more visible on land; these include loud vocalizations, head bobbing, and male-male combat. Characteristic tortoise behaviors are discussed throughout this book in relation to physiology, reproduction, and habitat use.

Inside the Shell: Visceral Architecture

The deep or high-arched shells of tortoises provide a lot of space for the internal organs (Fig. 3.1). But the proportionately larger internal volume compared to most other vertebrates does not mean that the viscera are free to slosh around in the body cavity. In turtles, humans, and all other vertebrates, a connective tissue sheath—the peritoneum—encases the organs. The thin, flexible peritoneum holds the organs in place. Functionally, the peritoneum encloses three major organ groups: (1) the lungs (pleural); (2) the heart and its large arteries and veins (cardiac); and (3) the liver, digestive tube, kidneys, and gonads (visceral). Although called cavities, they are not

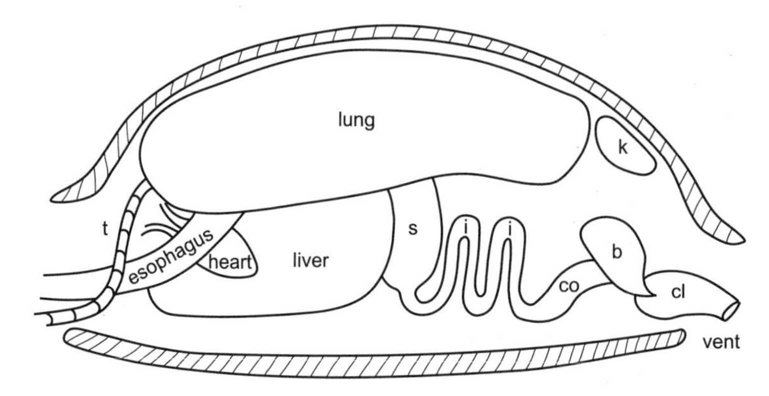

Figure 3.1
Sketch of the inside of a tortoise showing the general positions and relative sizes of the major organs (viscera) from a side view. Abbreviations: b = urinary bladder; cl = cloaca; co = colon or large intestine; i = small intestine; k = kidney; s = stomach; t = trachea. Sketch by G.R.Z.

open spaces because they are filled with the organs. In the case of the heart (pericardial cavity), a thin mantle of fluid allows the heart to beat freely.

Spongy Lungs

The chamber for the lungs—the pleural cavity—is the biggest cavity and fully occupied by the lungs in tortoises. It extends dorsally (along the back) from near the front of the shell to the shoulder girdle muscles to near the pelvic musculature, and the lungs' anterior halves adhere firmly to the underside of the carapace. Because tortoises are hidden-necked (Cryptodira) turtles (Appendix Table A.1)—their head and neck withdrawing straight back into the shell—most of the neck nestles back between the right and left lungs. Tortoise lungs are spongy, and when the head and legs retract, there is less space for the lungs. The reduced volume limits respiration but does not prevent breathing (see the Breathing and Oxygen Exchange section in this chapter).

Turtles' paired lungs are firmly enclosed within the pleural sheath (peritoneum) and can't collapse because (1) the peritoneum is fused to the underside of the carapace's surface membrane (periosteum), (2) the lungs sit tightly against the shoulder musculature, and (3) they're attached to the visceral peritoneum. The part of the pleural peritoneum where the lungs attach to the viscera is called the diaphragmatic membrane. It supports breathing by keeping the lungs open, but it's not a muscular pumping mechanism like the diaphragm of a mammal (see the Breathing and Oxygen Exchange section). The weight of the guts holds the diaphragmatic membrane down, keeping the lungs open. Picture how a saggy balloon would stay open if you put it in a box, glued it to the top inside, and weighted the bottom of it.

Ample evidence that tortoises float, sometimes long distances across oceans (Chap. 8), begs the question of whether their lungs explain their buoyancy. While lung volume seems like it should affect floating, tortoise lung volumes as a proportion of body weight are low (25% to 50%) compared to those of box turtles (51% to 93%), likely an adaptation to reduce water loss in the dry environments that tortoises inhabit. A turtle's ability to float depends upon its specific gravity: the volume (cm^3) divided by the weight (g) of an object. If its specific gravity is less than one (1.0), a turtle floats; if greater than one, it sinks. Eastern Box Turtles (*Terrapene carolina*) with their domed shells have specific gravities of 0.75–0.96 and float high in the water. A similar range of specific gravities can be expected for tortoises, particularly the giant tortoises. It's not just the lung volume, but also the total volume of the tortoise that confers buoyancy. Surprisingly, tortoise specific gravities are low, for example, 0.44 for Galápagos Tortoises, 0.37 for Mojave Desert Tortoises, and 0.2 for Leopard Tortoises.

Three-Chambered Heart

The chamber for the heart—the cardiac cavity—lies right above the plastron in the front third of the shell, tucked behind the front legs. The esophagus and trachea run overtop on their way to the stomach and lungs, respectively. The cardiac cavity abuts the sizeable bilobed liver and the front portion of the stomach.

On the outside, mammal and reptile hearts look similar, with two thin-walled upper chambers (atria) in the front separated by three large blood vessels emerging from a muscular chamber (ventricle) behind. Internally, in both groups the atria are separated by a wall of tissue, and each one empties independently into the ventricle. Like most reptiles, tortoises have just one ventricle chamber; however, their single ventricle is structured into chambers by three distinct depressions separated by muscular ridges, which maintains some separation between well-oxygenated and poorly oxygenated blood. From one chamber (the cavum pulmonale), low-oxygen blood flows into the pulmonary artery to the lungs. With a contraction of another chamber above (the cavum arteriosum), the muscular ridge blocks blood from the cavum pulmonale and directs the oxygenated blood into the body.

These chambers align with atrial openings and the entrances to the main exits of the large blood vessels (left and right aorta, pulmonary artery). Each of the main arteries (aorta) carrying blood away from the heart curves up and then back along the turtle before merging with its left-right pair into a single dorsal aorta that extends toward the tail. Before they merge, the right aorta sends a large (brachiocephalic) artery to supply blood to the vessels of the front legs and head. The left aorta produces three large branches before joining the right aorta. Those branches supply the viscera from the stomach rearward. The dorsal aorta supplies the various organs on the top of the body wall, for example, kidneys, gonads, and eventually the hind legs, girdle, and tail.

Vital Viscera

The remainder of the stomach, plus all the intestines, extends through the body cavity to the tail end, adhering to the diaphragmatic membrane above. Tortoises have large bilobed (two lobes) bladders for water regulation, occupying the rear of the body cavity immediately in front of the pelvic musculature. In turtles, the kidneys, gonads, and their ducts lie just beneath and attach to the underside of the carapace in the rear third of the body cavity, above the folded small intestines. The gonads are attached to the body by membranes; these attachments are flexible to allow for their enlargement during the reproductive season. The lungs, kidneys, and adrenals are more firmly attached.

The small intestine exits the stomach and lies in a series of folds in the middle of the body directly above the plastron. Tortoises that strictly eat plants (herbivorous) have longer intestines to ensure a longer passage time for digestion and absorption of nutrients from the food bolus, whereas tortoises that include more animal matter in their diets (omnivorous) have shorter intestines because these food items digest faster. The last part of the tortoise intestine is enlarged into a tubular colon, the main region for the reabsorption of water from food bolus. The colon empties directly through the anus and to the outside through a vent.

The Everything Vent

Underneath a turtle's tail is the vent, a slit opening that is controlled by a large sphincter muscle. In females, the vent is just under the edge of the carapace, whereas in males it lies beyond the edge of the carapace to accommodate the exit of the penis. The vent is the exit from a multipurpose cavity called the cloaca ("sewer" in Latin) shared by the digestive, excretory, and reproductive systems. The cloaca is a long, flexible chamber that starts just behind the pelvis and extends to its exit, the vent. Although the reproductive and excretory ducts are in separate left-right pairs, the bilobed bladder has a single duct (urethra) and opening. Each one empties into the cloaca through a separate duct opening. The food waste from digestion (feces), the nitrogenous (containing nitrogen) waste from the kidneys (urine), and the eggs or sperm pass through the cloaca via the penis's seminal groove to the outside through the vent.

Breathing and Oxygen Exchange

Until physiological testing in the 1960s, most biologists thought that turtles inhaled and exhaled by pumping their throats (gular pumping), as they make regular up and down movements of the throat as they breathe. But it turns out these movements are associated only with drawing air in and out of their nostrils into the olfactory chamber for smelling, not for respiration. Tortoises do not have a secondary palate, and the olfactory chamber lies at the front of the roof of the mouth, immediately adjacent to the internal opening of the nasal passage.

Breathing Mechanism

Tortoises (and other turtles) breathe with an adapted subset of the muscles used by other vertebrates. The vertebrate muscular mechanism for inhaling and exhaling is an ancient one that engages the abdominal muscles in expanding and contracting the chest cavity (through the rib cage) to draw air in and out of the lungs. Turtles lost most of these abdominal muscles as they evolved a rigid chest cavity stabilized by the carapace and plastron (Chap. 8), but two pairs remain: the obliquus abdominus and transversus abdominus. Their retention was necessary for powering inhalation and exhalation. These two pairs attach to the "diaphragmatic septum" (the horizontal barrier between the lungs and other internal organs).

For breathing in (inhalation), the abdominal obliquus (which is connected to the skin of the hind leg) contracts and draws the septum down, creating negative pressure that draws air into the lungs. For breathing out (exhalation), the abdominal transversus (connected to rear of the carapace) contracts, allowing the septum and viscera to rise and compress the lungs, driving air out. Tortoise front legs also move during breathing, rotating outward with exhalation and inward with inhalation. Their movements resulting from the alternate contraction of two pectoral girdle muscles contribute to both inhalation and exhalation but are not critical for oxygen exchange, as shown by captive tortoises commonly wedging themselves headfirst into corners, thereby preventing leg movements but continuing to breathe regularly.

Breathing Pauses

Turtles, including tortoises, are episodic breathers. They take one or more breaths and then pause, effectively stopping gas exchange with the environment. Some aquatic turtles may take many breaths and then pause for as long as a half hour; tortoises tend to take a single strong breath, followed by a few weaker breaths, and then stop for no more than a minute. This tortoise "singlet breathing" pattern contrasts with the variable patterns in aquatic turtles. Our knowledge of tortoise breathing is based on just a few rigorous observational studies, however, and the data on respiration in aquatic species derive from experimentation on fewer than a dozen species and principally from two species.

Regardless, the episodic breathing is likely a trait that evolved long ago for staying underwater in aquatic environments, an adaptation that tortoises retained when they adopted terrestrial lifestyles. Episodic breathing could still prove useful in some contexts, for example, for surviving long periods in burrows or other tight spaces with little oxygen, such as observed in Red-footed Tortoises (*Chelonoidis carbonarius*). Some have claimed that tortoises stop breathing when fully withdrawn into their shells; however, respiration more likely continues at a lower level, restricted by the reduced mobility of the viscera that would otherwise cause expansion and contraction of the lungs.

Oxygen Exchange

Turtles as a group, including tortoises, require less oxygen and have lower metabolic rates than most other reptiles. Their lower metabolism is reflected in tolerance of low-oxygen (hypoxic) environments when hibernating or estivating (Chap. 6). They are also able to survive for long periods with little or no food and water. These tolerances are associated with respiration, both with the movement of the air in and out of the lungs and the cellular physiology of extracting oxygen (O_2) and removing carbon dioxide (CO_2).

Turtles breathe with their mouths closed. Air enters through the nares (nostrils on their beak) straight into the mouth cavity and then continues into the windpipe (trachea). As in mammals, the trachea divides into two large tubes (the bronchi), one for each lung. The bronchi then branch out into multiple chambers for the exchange of O_2 and CO_2. Though the branching pattern is simpler than in mammals, turtles have an impressive network of branches (bronchioles) that lead to small respiratory tubes (alveoli) where gas exchange occurs.

The control center for breathing resides in the brain stem for all vertebrates. Increased carbon dioxide (CO_2) levels in the bloodstream stimulate nerves in the respiratory center of the brain (medulla and pons) to increase the speed and depth of breathing. Decreased blood CO_2 levels cause the breathing rate and depth to decrease. Blood oxygen (O_2) levels only become relevant if they get perilously low. But turtles have an intermittent nerve discharge pattern, matching their on-off breathing cycles. When the partial pressure of oxygen (pO_2)—a measure of concentration—gets to a certain high threshold, it shuts breathing off. As blood oxygen level falls, pCO_2 rises. Then, at a certain threshold of pCO_2, the brain stem neural mechanism is triggered, and breathing resumes. Tortoises' tissues, blood, and lungs store carbon dioxide during periods suspended breathing—apnea—which may help sustain them for longer.

Turtles overall better tolerate high pCO_2 levels in the absence of oxygen (anoxia). They can easily switch to oxygen-free (anaerobic) respiration and continue to produce energy for cell metabolism under low-oxygen conditions. Anaerobic respiration causes a buildup of lactic acid, causing dangerous acidic conditions that limit anaerobic respiration in most vertebrates. But turtles mobilize calcium and other ions from their bony shells that buffer the lactic acid, and store excess lactic acid in their tissues until oxygen becomes available again.

Blood Circulation

Blood Cells

At least in some species (e.g., Geometric Tortoises, *Psammobates geometricus*), the larger females have smaller red blood cells (erythrocytes) than males or juveniles, and adults have more stretched-out (elongated) erythrocytes than juveniles. Smaller, elongated red blood cells have greater surface to volume ratios, and thus more surface exposed for exchange of oxygen. Animals like birds and mammals with high metabolic needs for oxygen have smaller, more elongated red blood cells than reptiles. The smaller, more elongated red blood cells of female tortoises (relative to males or juveniles) may reflect their higher metabolic demands when they produce and lay eggs. Evidence that female Mojave Desert Tortoises (*Gopherus agassizii*) have more seasonal variation in red blood cell mass than males, with highest mass in spring and summer, corroborates this hypothesis (Chap. 4).

Turtles have a special hemoglobin—the oxygen-carrying molecule in blood—that is distinct from other vertebrates. Turtle hemoglobin is unusually complex, with multiple forms that may allow them to carry oxygen at different temperatures. Hemoglobin also varies within tortoise species. A comparison of the hemoglobin of the Yellow-footed Tortoise (*Chelonoidis denticulatus*) and the Red-footed Tortoise (*C. carbonaria*) revealed that the latter had a higher oxygen affinity (held onto more oxygen, releasing less into the bloodstream). Although it's not yet known how the difference plays out in wild tortoises, hemoglobin compositions relate to a species' metabolic functions and specific environments.

Going with the Flow

Reptiles in general have a more flexible circulatory system than other vertebrates, adjusting blood flow into their vessels differentially depending on the physiological needs of different parts of the body. For example, when the blood is fully oxygenated or a turtle is holding its breath, flow into the pulmonary artery is cut off; when more oxygen is needed, the pulmonary artery flow is restored. Bypassing the lungs can also be useful when a turtle needs to heat up by sending warm blood from the skin to the organs deep inside.

Eating and Digestion

Only the terrestrial North American box turtles (*Terrapene* spp.) and a few of their semiterrestrial relatives can eat on land. Aquatic and semiaquatic turtles have small tongues and few oral glands. They rely on water to move food into their mouths and back into the pharynx for swallowing. In contrast, box turtles and tortoises have well-developed tongues, which they use to manipulate food in the mouth and move it rearward into the pharynx. Numerous oral salivary glands help lubricate their food so it slides down easily. Box turtles and aquatic turtles use their jaws to capture food. Most tortoises, in contrast, use their tongues to capture food.

Tonguing the Food

A tortoise's tongue is large and muscular, filling the floor of its mouth. The upper surface of the tongue is a carpet of folds and villi with numerous pore openings for the secretion of mucus. The front quarter secretes an especially sticky mucus. To capture food, a tortoise usually approaches the item straight on, extending its neck, tilting its head slightly to see the food, opening its mouth, and extending its tongue to touch the food. With their eyes positioned on the sides of their head, once tortoises extend their heads toward a food item, they can no longer see it. Touching it with their tongues likely identifies the item and assesses whether its edible.

A tortoise presses its tongue onto the food to get the item to stick, and then retracts it to draw the food in. This "lingual prehension" (grasping with the tongue) is common to all tortoises except the North American genus *Gopherus* and Asian genus *Manouria,* which use their tongues only to manipulate food once it's inside their mouths. Their other primitive characteristics suggest that these exceptions to lingual feeding represent an older point on an evolutionary continuum. Aquatic ancestors presumably fed entirely underwater, as do most aquatic turtles today. Further along the continuum are aquatic turtles, such as mud and musk turtles that venture onto land and grasp food with their jaws and then drag it to the water for ingestion. Next are *Gopherus* and *Manouria* that grasp with jaws but feed on land. The remaining tortoises have the specialized fleshy, muscular tongues. Coordinating the muscles to extend the head, open the jaw, protrude the tongue, withdraw the tongue, and engage the jaw in chewing is a complex muscular choreography indicative of an advanced feeding mode.

Sensory Jaws

Tortoises and other turtles have powerful jaw muscles; the jaw adductor muscles are on a pulley system attached to their quadrate bone. The arrangement allows for longer muscle fibers to close the jaw vertically in an extra-strong bite. So, despite their relatively small skulls, tortoises can clamp down hard—an ability useful for cutting and crushing dense plant foods (Plate 11). The keratinous jaw coverings (tomia) lack nerve endings, but the dermal layer beneath has sensory receptors. These sensory corpuscles (little cells) are mechanoreceptors that signal to the tortoise the position and hardness of an object in its jaws. This mechanoreception is comparable to sensing pressure on our fingernails (which have no receptors of their own). Although lacking teeth, the tomia have sharp edges, serrated in some (e.g., Leopard Tortoise), allowing the cutting of the food into manageable sizes for swallowing.

Swallowing the Bolus

A tortoise opens and closes its mouth, gaping and chomping, several times as its tongue moves the item back into the pharynx for swallowing. The food may be compressed but not chopped up. The tongue, the hyoid, and muscles in the floor of the mouth and pharynx contract in a roughly wavelike manner to move the food into the esophagus. Once in the esophagus, the flexure and contraction of the neck muscles in something that resembles peristalsis (waves of contractions) move the bolus (ball of food) into the stomach. Although the stomach is largely a curved tube lacking the separate chambers of many herbivorous mammals, its walls are regionalized by different glands, providing digestive fluids to break the food down.

Digesting Plants

The food bolus continues through the stomach into the small intestine for further enzymatic digestion and absorption of mineral and nutritional components. The chemical processing of the bolus is almost complete as it moves into the large intestine. In addition to the final digestive activity in the colon, water is extracted from the bolus, moved by the circulatory system into the body, and eventually stored in the urinary bladder. This water retrieval is an essential process for keeping the right water balance (homeostasis) and a major adaptive feature that allows many tortoise species to survive in arid environments with little and infrequent access to free water. Water extraction greatly reduces the volume of waste.

Tortoise hindguts (large intestines) play an essential role in the final part of the journey of the bolus through the digestive system. Animal matter is the easiest food to digest. Fruit and young leaves also break down readily, but older and especially dry plant matter is slow to digest. Indeed, most dry vegetation passes through the digestive tract undigested; undigested plant pieces are visible in the feces of desert species of tortoise. A tortoise hindgut hosts a variety of microbes (fungi, bacteria, and protozoa) that break down the major components of plants that would be otherwise indigestible—the carbohydrates cellulose and hemicellulose. This gut microbiome of tortoises and other herbivores, such as rabbits and elephants, further ferments the smaller molecules, which converts them into fatty acids and other molecules that can be absorbed as nutrition while yielding energy.

The need for a proper gut microbiome in herbivores has also given rise to coprophagy (eating feces). Mojave Desert Tortoise juveniles eat the feces of adults to inoculate themselves with microbes for their increasingly vegetarian diets. Biologists have not confirmed that the young of other tortoise species are coprophagous, but it seems likely. There are numerous reports of juvenile and adult tortoises of many species eating the feces of mammals.

Bolus Passage

Passage time of food through the gut is variable, not fixed. A study of Leopard Tortoises (*Stigmochelys pardalis*) exposed to a diet of either tomatoes or lucerne, a ground cover plant, had faster gut transit times with the tomatoes but retained more water and energy from the lucerne. The ability to regulate the transit time of materials through the gut allows tortoises to take advantage of fluctuating food and water resources. Red-footed and Yellow-footed Tortoises demonstrated flexible digestive processes depending on the food. A diet of fruits with tough cell walls caused them to do little cell wall fermentation and instead extract the cell contents; in a diet of leaves, in contrast, the cell walls were digested. Time in the digestive tract also relates to tortoise access to heat. In captive Galápagos Giant Tortoises (*C. nigra*), the time that food spends in the gut (in one study ranging from 6 to 28 days) is inversely correlated with ambient temperature.

Because tortoises digest relatively slowly, they spend relatively little time eating food, with long, inactive gaps between feeding bouts. A study of Speke's Hinge-back Tortoise (*Kinixys speckii*) reported that only 6.6% of their time was spent foraging. Because digestion can take days, it might be important for a tortoise to select high-quality and easily digestible food (Plate 5) rather than fill itself with whatever it comes across. Indeed, the effort tortoises put into feeding matches the quality and availability of food. If seasonal rains have stimulated new growth of annual plants dubbed "tortoise turf," tortoises such as the Aldabra Giant Tortoise (*Aldabrachelys gigantea*) will begin to graze at first light and continue until the heat of the sun drives them to seek shade to avoid overheating. Similarly, Berlandier's Tortoises (*Gopherus berlandieri*) will gorge on the fruits ("tunas") of Prickly Pear (*Opuntia* sp.) during the ripening season.

Tortoises also eat sand, small stones, or bone fragments, a behavior seen in some birds and fish. Although these abrasive materials may be retained as gastroliths (grinding elements) in the stomachs of other vertebrates to assist with the breakdown of plant cell walls, this function has not been verified for tortoises. Eating soil, stones, and bones likely helps maintain gut pH, control intestinal parasites, file down tortoise beaks, and obtain minerals such as calcium for female egg production (Plate 6).

Water and Hydration

Water and body hydration are essential for a tortoise's survival. Regular access to water is limited for most species owing to their dry environments. Even most tropical forest species, such as Burmese Star Tortoises (*Geochelone platynota*), live in areas with seasonal rainfall that is not continuous throughout the year. Myanmar and much of India and Southeast Asia are dominated by a monsoonal climate, which results in several to as many as nine months of seasonal drought. Whatever the environment, tortoises have developed behavioral, dietary, morphological, and physiological mechanisms to maintain adequate body hydration.

Drinking

The simplest way to maintain hydration is to drink water. Tortoises drink, but in ways as unique as their body plan. Short tongues prevent tortoises from lapping up water like a dog or cat. Also, the forward plastron projection ("gular") on some male tortoise species (Chap. 4) is in the way of easy downward access to water unless the neck and head are fully extended.

Tortoises usually approach water in their slow, methodical gait. At the water's edge, the front portion of the plastron is lowered to the ground, the forelegs sprawled diagonally forward (giraffe style), and the head and neck extended and angled downward, so the head is partially submerged with the nostrils underwater (Fig. 3.2). Then the throat and bottom of the mouth begin a slow upward and downward movement (buccal pumping).

Observers have assumed that water enters a tortoise's buccal (mouth) cavity through the submerged nostrils. Another hypothesis is that its mouth opens just a

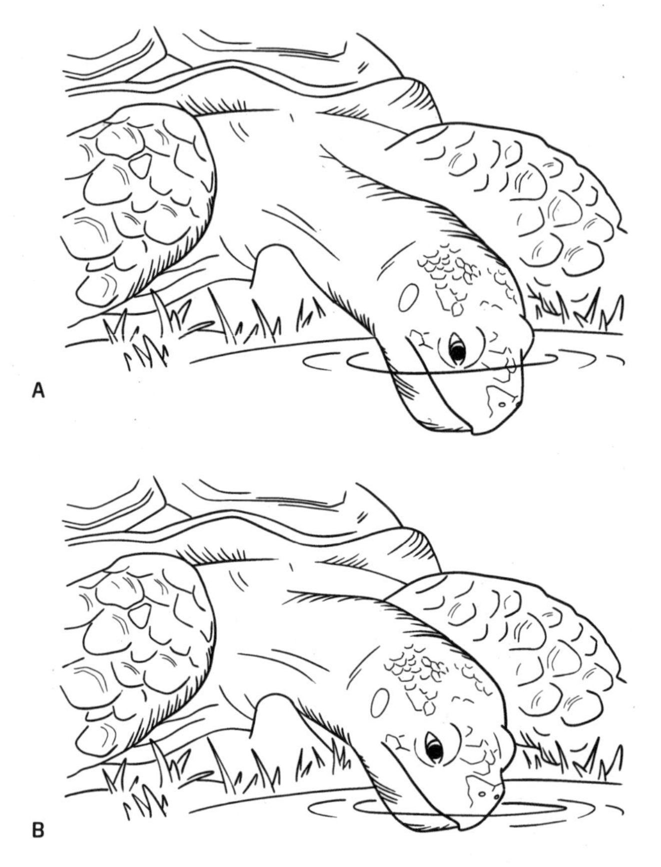

Figure 3.2
Typical head position of a tortoise drinking water. **(A)** Typical head immersion if the water source is deep enough. **(B)** Head position in a shallower puddle. If the water is just a shallow puddle or thin sheet preventing emersion, the head is held vertically with only the tip of the snout and nostrils in the water (see also Plate 8). From sketches by G.R.Z.

bit, making an elongate slit several times wider than the nostril openings. Bringing water in through its mouth could prevent flooding of a tortoise's nasal tissue with each drinking pulse. Also, the duration and size of the throat pulses during drinking suggest water entering in more volume than would be possible through the nostrils.

One tortoise group—the Aldabra Giant Tortoises—definitely drink through their nostrils to get water from small depressions in coral rocks (Plate 8), the only available fresh water on Aldabra Atoll. Researchers noticed that Aldabra Tortoises inserted only the tip of their snouts in the water. When offered water in a tablespoon, the Aldabra Tortoises "inhaled" it. Subsequent examination of the nasal anatomy of the Aldabra Tortoises and Mascarene species revealed an olfactory chamber set off from the anterior nasal cavity and the buccal cavity, and furthermore with a valvular flap that prevents water entry. Other species from Radiated Tortoises to Mojave Desert Tortoises have been observed drinking from shallow puddles of water and even pressing their noses into moist soil, seemingly to extract water. The nasal anatomy of most other tortoise species has yet to be examined.

Hydration Innovation

Most tortoises live in arid or semiarid habitats where water is scarce, and some novel ways of accessing drinking water have evolved. During rainfall, the Tent Tortoise (*Psammobates tentorius*) elevates the rear of the shell, allowing water to follow the scute sutures down to the front marginals. From the marginals, the water drips onto the forelegs, from where tortoises lick the water with their tongues. During rainfall, some Gopher Tortoises will move to the entrance of the burrow, turns sideways, and extend their upslope forelegs tightly pressed against the soil, which creates a dam and a puddle of water from which they drink.

The hotter the weather, the more tortoises salivate (much like dogs) to cool off through evaporative water loss, and thus the higher the need for behavioral and physiological means to stave off dehydration. One outstanding behavioral feature shared by tortoises is the mass exploitation of the initial rains that end the dry season. Mojave Desert Tortoises create shallow bowls in the hard soils of their habitats. When the rains begin, they emerge from their burrows and "race" to these bowls. In southwestern Madagascar, nearly 100 Radiated Tortoises (*Astrochelys radiata*) were observed concentrated in a creek bed as the first rainstorm of the wet season began. The tortoises had slowly emerged from the dry scrub and descended into the creek bed in anticipation of growing puddles of water.

Surviving Drought

Broadly, tortoise movements are patterned on water availability. For example, the activity patterns of Mojave Desert Tortoises shift year to year depending on the amount of rainfall. In especially dry years, they significantly restrict their aboveground activities, spending more time hunkered down in burrows where they're less vulnerable to evaporation. Drought limits the growth of water-rich annual plants, shifting the cost to benefit ratio for spending time outside the burrow foraging versus staying in the burrow safer from predators.

Tortoise feces, especially of species living in desert environments, are dry because tortoise digestive systems actively extract water from the bolus and transfer it via the bloodstream to the urinary bladders. During times of drought, tortoises maintain a safe osmotic balance for their cells by periodically releasing water from the bladder into their blood plasma. English comparative physiologist Robert Townson first proposed that tortoise bladders serve as water reservoirs, a feature exploited by thirsty sailors and explorers visiting the Galápagos Islands (Chap. 9).

Tortoises excrete their nitrogenous waste as uric acid crystals and urates that precipitate out of bladder water. As a dry spell persists, the concentration of metabolic waste increases to toxic levels, but the walls of the bladder retain the concentrated urine and prevent its transfer into the bloodstream. During an extreme drought year, the urine of Mojave Desert Tortoises had some of the highest level of osmolality (particle concentration) known for terrestrial reptiles. As drought continued, even the bloodstream osmolality spiked to levels that would normally be toxic, showing that tortoises can tolerate temporary deviations from homeostatic (balanced) osmotic conditions. When rains return, a tortoise voids its bladder of the toxic urine and drinks copiously to rehydrate its body with fresh water. If sufficient water is available during their "re-tanking," a tortoise can imbibe anywhere from 10% to 28% of its body weight.

Homeostasis

Tortoises adjust a suite of digestive parameters as needed to maintain suitable water balance. Leopard Tortoises show digestive flexibility in captivity that suggests they could adjust food intake, food transit rates, and urine osmolality in the wild. Still, hydration is challenging. Tortoises that live in more seasonal climates exhibit more extreme fluctuations in body condition in tandem with changes in vegetation and water availability. Thus Mojave Desert Tortoises (California) have higher fluctuations in body condition than Gopher Tortoises (Florida) because of the more stable rainfall of the latter's habitat. Even if total rainfall is the same, timing matters. During a year when most rainfall was concentrated in the winter, Speckled Dwarf Tortoises (*Chersobius signatus*) suffered poorer body condition than when rainfall occurred throughout the year.

Because they are smaller, juvenile tortoises may have less capacity than adults to rely on metabolic osmoregulation to stay hydrated. With high surface area to volume ratios, they're more vulnerable to evaporation of water from their bodies and must behave in ways that reduce it. Juvenile Mojave Desert Tortoises ameliorate high evaporative water loss by drinking rainfall, withdrawing into their shells to reduce exposed surfaces, and taking refuge in burrows (Chap. 6).

As ongoing climate change continues to alter rainfall patterns, smaller tortoises and species living in more seasonal climates might be at higher risk. Their time-tested adaptations to maintain hydration may prove fundamental to tortoise survival during major climatic fluctuations.

Temperature Regulation

Hot, cold! Wet, dry! All organisms live and reproduce within a specific range of environmental conditions. As noted above, most tortoise species live in areas of limited water availability for at least part of each year. They've evolved physiological and behavioral characteristics that allow them to survive and reproduce in these habitats. Similarly, each tortoise species occupies habitats with a particular temperature regime. On a worldwide basis, they live in habitats bounded in the north by the 0°C isotherm (roughly southern Canada–southern Europe) and in the south by the 10°C isotherm (South America's Cape Horn; Table 3.1).

Ectotherms

Tortoises are ectotherms, which depend on the environment to provide them with heat to keep their bodies within a range of temperatures suitable for normal cell function. If cells get excessively hot, cellular components become denatured (the conformation of molecules changed). If cells get too cold, chemical activities become too slow to be effective, and if cells freeze, water crystals form and mechanically rupture the cell membranes, killing the cells. Tortoises avoid temperature disruption of cellular functions through a variety of morphological, physiological, and behavioral adaptations.

Tortoise metabolic rates generally scale with temperature, increasing as body temperature rises and decreasing as body temperature falls. Most tortoises have a preferred temperature range of 30°C to 35°C for their daily activities. Many can remain active at lower temperatures, for example, 24°C–29°C, although they seek thermal refuges as temperatures drop into the teens. Below about 10°C, their low metabolisms make them immobile. Even though a few species, such as the Steppe Tortoise (*Testudo horsfieldii*) and Berlandier's Tortoise, can survive freezing (likely

Table 3.1.
Temperature Equivalency of Degrees Celsius to Fahrenheit

Celsius (°)	Fahrenheit (°)	Celsius (°)	Fahrenheit (°)
0	32	25	77
5	42	30	86
10	50	35	95
15	59	40	104
20	68	45	113

via a super-cooling mechanism that keeps ice crystals from forming in blood and other tissues), most tortoises seek refuges that keep their body temperatures above freezing during the winter.

Heat Stress

Heat is more dangerous than cold because high temperatures damage cellular processes and structures. There are species-specific high temperature limits beyond which tortoises can't survive. As ambient temperatures approach and exceed the preferred range for a particular species, they first respond by limiting their movements. Speke's Hinge-back Tortoises, for example, become inactive when temperatures exceed 29°C (84.2°F). Tortoises usually show heat stress at 37°C–38°C. At 38.4°C, Speke's Hinge-backs start to salivate to prompt evaporative cooling, as do the Leopard Tortoises with which they overlap in range. The critical thermal maximum or lethal temperature that kills most tortoises is 42°C–44°C.

The shell would presumably be a passive structure in temperature regulation, but it's not. Both the bone and the epidermal scutes act as heat shields to retard overheating in direct sunlight. Though the bone may act as an insulation layer, the scutes resist heating and reflect the sun's heating rays. Exactly how the scutes serve this reflective function is not clear. When the shells of living Mojave Desert Tortoises were experimentally painted with either reflective or absorptive paint (that could be easily removed), body core temperatures of the tortoises did not differ between painted and untreated samples. The role of the scutes in thermoregulation may therefore derive from their structure (Chap. 2), not their color.

Seeking Shelter

The main way tortoises thermoregulate is through behavior. Tortoises seek shelter to avoid extreme temperatures. All the smaller and many of the medium-sized tortoise species dig shallow cavities (pallets) that are typically adjacent to a tree, shrub, or large rock rather than out in the open. Except for the Berlandier's Tortoise, all North American Gopher Tortoises dig burrows of a few meters or more. These burrows have more stable temperature and humidity that allow a tortoise to escape the extremes of summer and winter. The African Spurred Tortoise (*Centrochelys sulcata*) also digs deep burrows, enabling it to survive in the African Sahel (Chap. 6).

Most species of tortoise also depend in one manner or the other on vegetation for shade. Egyptian Tortoises (*Testudo kleinmanni*; Plate 7), for example, use different types of vegetation for changing seasonal needs. In the fall, winter, and spring, they take shelter under small- to moderate-size shrubs less than 1 m in diameter. The shrubs provide both shade and concealment from predators. In the summer, however, when high temperatures force Egyptian Tortoises to estivate (become dormant), they use shrubs more than a meter in diameter as thermal refuges.

Daily Schedules

Most tortoises are strongly diurnal, their activities restricted to daytime. One factor driving diurnal activity is presumably avoiding predators, to which tortoises are more vulnerable at night because of their poor nocturnal vision and slower response times. If the temperature is suitable, they'll become active at first light of the day. But survival in areas with extreme daytime temperatures may require nocturnal habits. Bolson Tortoises (*Gopherus flavomarginatus*) live in Chihuahuan desert habitat, and during the hottest time of year (when even deep soil temperatures reach ~34°C) they revert to nocturnal feeding. When they emerge at night from their burrows, their body temperatures are about 28°C–30°C, and because of their large body mass, they lose heat slowly while grazing. There's some evidence that Impressed Tortoises (*Manouria impressa*) are also active at night. They spend most of the day hidden under leaf litter on the forest floor.

Even within a species, daily schedules may vary depending on where individuals live. The daily hazards of sun exposure are high on Aldabra Atoll's main island Grand Terre where the main food source of herbs, grasses, and sedges is in open areas (Plate 9). Tortoises must graze early and then compete for shade under the scarce trees (Plate 10) for most of the day before returning to graze in late afternoon. In contrast, in the wooded areas where sun exposure is reduced, tortoises can graze all day.

Seasonal Shifts

Most tortoise species live in areas with marked seasonal cycles of temperature and rainfall regimes. Seasonal shifts in activity to cope with the changes are widespread, from the European *Testudo* species to the multiple genera (e.g., *Homopus*, *Psammobates*) of the southern African scrublands and savannahs. For example, Gopher Tortoises live in a seasonal environment (southeastern North America) with winter temperatures that drop near freezing and midday summer temperatures that regularly exceed 32°C. These tortoises hibernate in their burrows once daily temperatures fall below ~14°C, although they may come to the mouths of their burrows on warm winter days (~18°C). In the spring and fall when temperatures at midday are ~22°C–28°C, they're active at midday. In summer with temperatures ~30°C–38°C, they shift to a bimodal activity cycle of early mornings and late afternoons.

Seasonality involves not only temperature, but also rainfall. The miniature Spider Tortoises (*Pyxis arachnoides*) live in one of the driest habitats, the thornscrub of Madagascar. They estivate (remain dormant) under the surface leaf litter beneath scrubs and trees from April through November, becoming active only in the wet season from December through March.

Thermal Inertia

The bigger the tortoise, the more slowly it heats or cools because of its lower surface area to volume ratio; tortoise surface area increases roughly at a 2:3 ratio to volume. As a rule, smaller tortoises—because they heat up faster—have lower maximum temperature tolerances. Giant tortoises, such as those on Aldabra, can graze for longer, but with the drawback that (because of thermal inertia) they may continue to heat up even once they move into shade. If a giant tortoise has not judged its heating rate properly, it can reach a critical or lethal body temperature even in the shade. Size, and therefore age, affects how tortoises adjust their behaviors to control heating and cooling rates.

In addition, juvenile tortoises are at high risk of predation (Chap. 5) and must select feeding times that balance the availability of their food (typically wet periods with young plants) and suitability of temperature (warm enough to allow rapid escape but not risk overheating their small bodies). A juvenile Gopher Tortoise was observed making a rapid escape into the burrow from an approaching predator (~45 m away), followed by a modest hiding time (>20 minutes).

Although tortoise metabolism is largely driven by outside temperature, some other factors come into play. Experiments with Angulate Tortoises (*Chersina angulata*) show a daily metabolic rhythm based on ambient light. Even when temperature remains constant, Angulate Tortoise metabolism follows a circadian rhythm: higher in the morning, decreasing during the day, and lowest at night, as evidenced by resting oxygen consumption.

Senses

Tortoises share basic sensory structures with (other) reptiles, mammals, and birds because of their shared vertebrate ancestry. Tortoises, however, have a unique balance of sensory strengths and weaknesses that relate to their mode of life. Tortoise sensory adaptations reflect the fact that they feed on land, a feature that evolved secondarily from turtles' aquatic feeding mode. As terrestrial herbivores who evade predators primarily by sheltering in their shells, tortoises have senses that are attuned to finding and evaluating plant food, connecting reproductively with mates, and navigating seasonally changing environments.

Binocular Color Vision

A tortoise's eyes are positioned on the side of its head for a broad field of view, an arrangement typical of animals that are not predators but must scan for them. Like many reptiles, tortoises have nictitating membranes, a set of inner eyelids that evolved as an adaptation for aquatic life but has been retained in turtles. Their complex eyes work together for binocular vision like ours, and they can see colors in the spectrum that is visible to humans. Tortoises likely have tetrachromatic color vision, however, based on four color receptors (in contrast to our trichromatic red-green-blue vision). Most reptiles have four color receptors, which extends the range of visible colors beyond the spectrum of 400–700 nm visible to humans. Because tortoises are largely diurnal, their retinas contain cone cells but no rods. While rods confer nighttime vision, cone cells are active at high light levels and responsible for color vision with high spatial acuity.

Favorite Colors

Although researchers do not fully understand the visual acuity of tortoises, lab experiments have revealed unique aspects of tortoise vision. For example, choice experiments show that tortoises prefer colors in the red to yellow range over blues or greens. The preference shows up in diverse tortoise families, including the Aldabra Giant Tortoises, Hermann's Tortoises (*T. hermanni*), and Red-footed Tortoises, suggesting that it may be a trait of all tortoises. Reds, oranges, and yellows are typical of edible fruits and flowers in tortoise habitats and get their yellowish color from pigments called carotenoids that play important biological roles, such as enhancing the immune system. Animals cannot synthesize carotenoids and thus must get them through food. Tortoise responsiveness to yellowish colors may be an adaptation for finding high-carotenoid-content plants. Male and female Hermann's Tortoises have proven equally capable of discriminating between colors, suggesting that the ability relates to food selection rather than, for example, sexual signaling.

Navigation

Tortoises' sense of sight also serves in navigation and homing relative to their home ranges. Gopher Tortoises can navigate back to burrows if displaced up to 200 m away unless their eyes are covered. That said, food selection and orientation are not achieved by vision alone; like most terrestrial vertebrates, tortoises rely on a combination of visual and chemical cues (see the Sexy Smells section).

Controlled experiments conducted with Red-footed Tortoises illustrate the importance of eyesight, perhaps as their primary sense. In a challenge of navigating an eight-armed radial maze to a food reward, tortoises performed reliably better than chance in finding the reward, even if olfactory (smell) cues were removed. Coupled with studies of aquatic turtles demonstrating that in visual discrimination tasks they did not rely on smell, these findings suggest that all turtles may depend most heavily on vision.

The color, binocular vision of tortoises has also begged the question of whether they can recognize images and discriminate them from real objects. Seeing the difference—scientifically dubbed "representation insight"—is a complex task. Experiments with Red-footed Tortoises have not found them capable of representational insight; instead, they often confuse photographs of an object, such as a favored food, with the real object. These results show that they can see and interpret the visual image but not discriminate it from the food.

Depth Perception

Tortoises also have well-developed depth perception. Experiments in the early 20th century showed that whereas even blindfolded aquatic turtles would readily jump off a raised surface, terrestrial and semiaquatic turtles would not. Subsequent experiments with Red-footed Tortoises showed their avoidance of deep visual cliffs. Equally, three species of *Gopherus* (Berlandier's Tortoise, Gopher Tortoise, Mojave Desert Tortoise), when offered a choice of a deep or shallow cliff edge, avoided the deep side. The difficulty of interpreting the results of these "visual cliff" experiments is separating recognition of depth from the motivation for avoiding an edge and having a reason for avoiding it. For an aquatic species, falling off a log is an escape reaction into its habitat. For a tortoise to fall off an edge, even a small one, is dangerous because of the risk of landing on its back and expending energy to right itself (Chap. 2).

Great Sniffers

Along with vision, smell is the most significant sense of a tortoise. Turtles in general have a large suite of olfactory receptor genes, especially terrestrial turtles whose olfactory receptor genes may even outnumber those of mammals. The abundant genes reflect a strong sense of smell that comes from both their nasal chambers and what's called the Jacobson's organ (or the vomeronasal organ). A patch of cells on the roof of the mouth at the base of the nasal cavity, the Jacobson's organ detects chemicals used in communication between individuals of the same species, such as for reproduction. Whether the tortoise vomeronasal tissues constitute a true Jacobson's organ has been debated, especially given how much the tissues vary across species. Regardless, tortoises open their mouths slightly to detect odor particles in the air with those tissues. Generally, the vomeronasal tissues detect moisture-based odors, whereas the nasal olfactory system detects airborne odors.

Sexy Smells

The moisture-based odors include compounds released from several body locations, such as the mental glands and the cloaca. The mental glands (or "chin glands") are a pair of organs embedded in the skin just behind the tip of the lower jaw, known only from *Gopherus* and *Manouria*. Their presence in most aquatic turtles indicates that they may play a more prominent role underwater and have been retained by just these two more primitive groups of tortoises (Chap. 8).

Studies of Gopher Tortoises have shown that secretions of pheromones from the mental glands play a role in mate choice and recognition of individuals of the same species. Seasonally, the mental glands enlarge and stimulate social behavior during the mating season. Generally, male tortoises have well-developed chin glands that swell at sexual maturity and begin to emit a cocktail of proteins, lipids, and steroids during courtship and male-male combat. A male Mojave Desert Tortoise with larger chin glands has higher social status, and skin gland size is correlated with testosterone levels. The activity of chin glands also varies across tortoise species, presumably in tandem with their mating systems.

A more ubiquitous source of tortoise pheromones is the cloaca or opening to the reproductive and digestive tracts (Chap. 2). The cloacal odors are species-specific at least in some tortoises. Spur-thighed Tortoise (*Testudo graeca*) males may find females by their chemical signature. A field study showed males locating females hidden underneath bushes up to 40 m away without obvious visual or auditory cues. The males made beelines for hidden females and then sniffed female cloacae, suggesting that they responded to pheromone signals emitted by the females.

As they approach females, males bob their heads, a behavior closely associated with courtship (Chap. 4) and implicated in helping males sniff chemicals emitted by the females. Males are generally more responsive to female pheromones than the reverse, based on a study of Hermann's Tortoises showing that males can discriminate a sexually mature female from other tortoises by odor from her tail end. Males will also sniff a female's cloaca during courtship, presumably detecting chemical signals she emits that indicate health or reproductive condition.

It's Not All about Sex

Although smell by the Jacobson's organ appears to be the primary sense used in mating, airborne odors detected in a tortoise's nasal system might provide cues unrelated to reproduction. Those odors could, for example, include chemicals excreted by plants that indicate availability and, for fruits, ripeness. Airborne odors may also help detect predators such as minks or other mustelid animals with strong smells. They could also play a role in females identifying nesting sites with suitable soil conditions.

A Bit about Taste

Taste goes hand in hand with smell when it comes to food. Tortoises have a complex oral cavity with large glands and a well-developed tongue. Their tastebuds are not well studied, but at least in the Asian Giant Tortoise (*M. emys*), they're clustered mostly on the floor of the mouth near the front and along the top sides of the tongue. *Manouria* may not be typical of most tortoises, given its status as a basal tortoise (early from an evolutionary point of view; Chap. 8).

Hearing

Turtles and all other reptiles have an inner and a middle ear (but no outer ear other than a simple one in crocodiles). From the outside, a turtle ear looks like a round membrane-covered hole. The membrane is about twice the thickness of a piece of paper (0.25 mm for Hermann's Tortoises), although its diameter and thickness vary across species. Behind the membrane is the middle ear, which is basically an open space (tympanic cavity) that is completely closed off to the outside but connected to the inner ear by a bony rod (the columella). When the tympanum vibrates, the sounds travel down the bony rod through a large, liquid-filled sinus to the inner ear.

Low Tones

Tortoises perceive sounds in a much narrower range than humans. Experiments with Hermann's Tortoises demonstrated a perceptible range of 10 to 182 Hz (in contrast to our audible range of 20 to 20,000 Hz). So, tortoises hear only the deeper frequencies, including infrasound (10–20 Hz) not audible to us. Juvenile Gopher Tortoises fled into their burrows from as far as 45 m away when they could not yet see an approaching researcher (aka predator), suggesting they could detect vibrations. Because lower sound frequencies are only weakly absorbed by the environment, they travel farther, so a tortoise's hearing may be adapted to picking up sounds of predators, allowing sufficient time to escape.

Love Songs

Tortoise's receptivity to low-frequency sounds likely plays a role in reproduction. It may allow females to evaluate male calls during courtship and mounting as potential indicators of male fitness (Plates 13 and 14). For example, vocalizations by Spur-thighed Tortoises change according to male body condition and are thought to have evolved through a female preference for healthier males. And female Hermann's Tortoises specifically prefer higher-pitched, fast calls when given a choice in an experimental setting. High-pitched calls correlate with smaller male body size and higher aerobic capacities, suggesting that females may select for certain qualities that would confer fitness to their offspring. The bottom line is that sound and hearing are an integral part of tortoise reproduction, and likely mate choice (Chap. 4).

Tortoise Chatter?

Increasingly, biologists are recognizing that vocalizations are an important mode of communication across many types of turtles, including aquatic species. The functions of vocalizations may extend well beyond reproductive and foraging behavior into aspects of social behavior that are not fully understood. In Gopher Tortoises, researchers have observed "conversations" going on in which tortoise are seemingly calling to each other from adjacent burrows. Their low-frequency vocalizations (3–40 Hz) may continue for up to 10 minutes. Juvenile Red-footed Tortoises vocalize as they forage, making chirps or clucks. Tranvancore Tortoises (*Indotestudo travancorica*) reportedly chorus at night. Tortoises likely use sound for a range of life history functions.

Touchy Subject

It may surprise some that tortoises can feel tactile stimulation of their shells. Studies of Spur-thighed and Hermann's Tortoises showed neural impulses in response to touches on the shell. Mechanoreceptor nerve cells in the surface layers receive and transmit signals when the shell vibrates. The receptive fields are localized and patterned on the scutes of the carapace (Chap. 2). The shell's sensitivity to touch is likely an important signal to both partners when males mount females during courtship and mating. Aldabra and Galápagos Tortoises assume a high stretch posture when legs or shell are "tickled" by birds ready to search their skin for ectoparasites, such as ticks (Plate 12).

Tortoises also have clusters of sensory corpuscles (endings of nerve cells) in their upper jawbone tissues toward the front of their mouths. They are tiny structures, just fractions of a centimeter (ranging from 60 to 150 μm in a Leopard Tortoise). Tortoise and other turtles' corpuscles probably play a role in monitoring the movements of the tomium (beak edge), like how corpuscles function in birds. How exactly the corpuscles function in tortoises is not well understood, partially because their structure is distinct. Each tortoise corpuscle has multiple nerve cells entering it (rather than a single nerve), and the corpuscles are buried within tissues such that they have no contact with the surface. Overall, the corpuscles of turtles are more complex than those of other scaled reptiles.

Cognition

Tortoise senses collectively contribute to their survival in the environments they inhabit, which has been a long success story until recently (Chap. 9). Although reptiles are not lauded for their learning abilities, experiments with tortoises in laboratory settings have demonstrated cognitive abilities in tasks that include memory and learning.

The Tortoise and the Rat

A Red-footed Tortoise trained in an eight-arm radial maze with olfactory cues removed (see the Vision section) could select the arms baited with food based on visual learning. Its spatial learning abilities were comparable to those of rats tested in equivalent mazes. Furthermore, if visual cues were unavailable, tortoises switched to a navigation strategy based on spatial memory and the application of decision-making that entailed consistently turning in one direction away from the arm just visited. It appears that tortoises use visual cues when available (most of the time, since they're diurnal foragers) but access cognitive abilities in other situations when needed.

Peer Learning

Laboratory experiments with Red-footed Tortoises also demonstrate abilities thought to be absent in reptiles. Observations of captive tortoises showed them following the gazes of other tortoises (conspecifics), which may provide information about food or other aspects of the environment. In addition, captive Red-footed Tortoises showed social learning, where a tortoise watching a conspecific solve a problem (going around a barrier to access food) later solved the same problem, whereas non-observer tortoises could not. This apparent ability of Red-footed Tortoises to learn from each other may be a social adaptation, although they and other tortoises are not considered to be social animals. Though they interact around courtship, mating, and in some cases home range assets such as burrows, tortoises are not thought to form permanent groups.

Kin Recognition

There is emerging laboratory evidence that tortoises can recognize relatives, or kin, despite their apparent lack of pair bonds or parental care of offspring (Chap. 5). A study of hatchlings from two species—Marginated (*Testudo marginata*) and Spur-thighed Tortoises—found evidence of spontaneous recognition of familiar tortoises and avoidance of unfamiliar tortoises. A hatchling encountering an unknown tortoise (of the same species) approached quickly, then retreated to a distance away, whereas hatchlings meeting known individuals paid little attention and maintained random spacing. Hatchling tortoises in the genus *Testudo* (Spur-thighed, Hermann's, Steppe, and Marginated) also demonstrate an attraction to dots arranged in a face-like pattern, which possibly stems from an ancient orientation toward conspecific faces (of relatives) from common ancestors of reptiles, birds, and mammals. Further experimentation with other tortoise species, coupled with observations of wild behavior, could illuminate the subtleties of tortoise cognition.

CHAPTER 4

Courtship to Nesting

Tortoise Reproduction

For tortoises, sex isn't easy. Indeed, it can be precarious with the male balancing nearly vertically over the rear of the female's shell, at risk of toppling over backward (Fig. 4.1). Imagine wearing a full suit of armor and trying to crawl up over a Volkswagen Beetle. Yet males persist and in fact pursue mating opportunities through noisy and aggressive courtship of females. Along with courtship comes male-male competition (Plate 13), which in some species comes to physical conflict. These obstacles are only the first difficulties of tortoise reproduction.

Figure 4.1
Portrait of female and male Ploughshare Tortoises (*Astrochelys yniphora*) copulating. Outline illustration after Pedrono (2008, Fig. 30)

All tortoises—indeed, all turtles—lay eggs, and with their high-volume shells, most tortoise species could easily house multiple large eggs. The eggs must be small enough to pass through the opening between the vertical ilia bones, however, as well as between the carapace and public bones (Chap. 2). This opening can't expand and is also the passageway for digestive and excretory ducts in all turtles; egg size is constrained by the diameter of pelvic openings. Then there's the feat of nesting on land, which occurs in all turtles. Tortoises may have the advantage of already living on land, but finding the right place for a nest that is safe from predators and has suitable soil and water conditions is nevertheless a specialized task. For reproductive success, female tortoises rely on repeat clutches of eggs over their long life spans while asserting a degree of control over when they nest relative to environmental conditions.

Clucks, Grunts, and Roars: Courtship

Tortoise males are attracted to females by chemical pheromone cues as well as visual cues. (Chap. 3) Once a male encounters a female, courtship is both active and potentially noisy (Fig. 4.2, Plate 14). The signaling system that guides courtship and mounting is based on the combination of visual, odor, and sound cues. Male tortoises are loud, vocalizing in croaks, hisses, and grunts as they court females. Egyptian Tortoise (*Testudo kleinmanni*) males, for example, make rasping and high-pitched calls that sound like pigeon squawks. The courtship sounds of Asian Giant Tortoises (*Manouria emys*) have been described as loud, rhythmic moaning that could be mistaken for a sick or dying mammal. Aldabra and Galápagos Giant Tortoises mating can be heard from a kilometer or more away

Casanova Style

Some aspects of courtship are common across tortoises, such as strutting, vocalizing, head movements, and mounting, but there are also species-specific courtship styles (Fig. 4.3). Distinct styles can be seen in three Mediterranean species of *Testudo*. Marginated Tortoise males swing their heads horizontally and bite the female hard before mounting her. Spur-thighed Tortoise (*T. graeca*) males bob their heads vertically and butt the female's carapace before mounting. Hermann's Tortoises (*T. hermanni*) mostly court from a mounted position, thrusting their tails back and forth during courtship (Plate 14). A specialized claw at the end of the exceptionally long tail stimulates the female. The wide, flared rear end of a Hermann's Tortoise carapace may provide more stability during the mounted courtship.

Figure 4.2
Male Red-footed Tortoise (*Chelonoidis carbonaria*) vocalizing while mounting a female.
Photograph by Bioparaguanero, Wikimedia CC-BY-SA-4.0

Rough Handling

Tortoise courtship from outside appearances can be rough, with a female withdrawing into her shell as the male circles around croaking, jerking his head around, and ramming or biting depending on the species. Females may get injured in the process, such as bite wounds on their legs, shells, or even around the reproductive openings (vents). Still, the roughness of courtship appears to have some bearing on female assessment of male quality. In Marginated Tortoises (*T. marginata*), the number of times a male bit and rammed a female was correlated positively with their mating success. In some cases, rough handling during courtship may be an artifact of denser

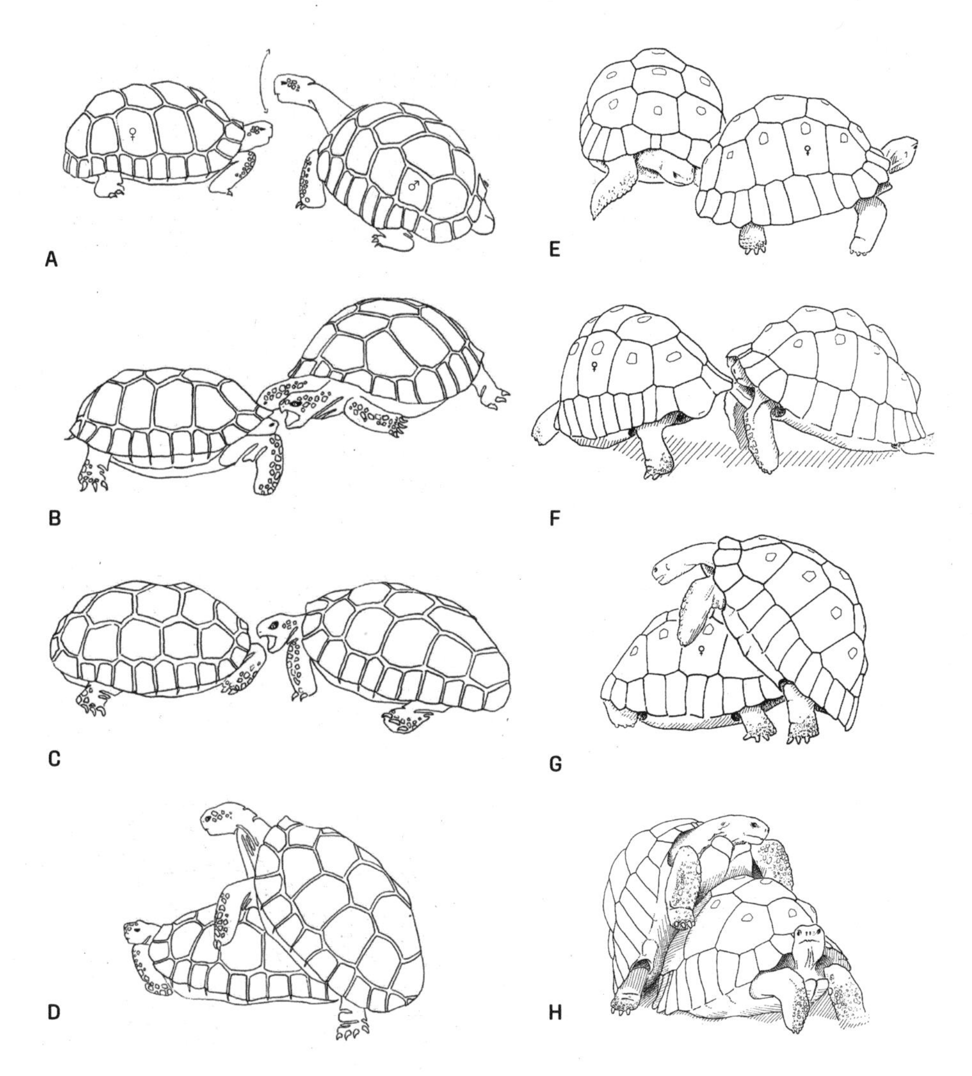

Figure 4.3
Courtship behavior. **(A–D)** Female and male Mojave Desert Tortoises (*Gopherus agassizii*), and **(E–H)** female and male Radiated Tortoises (*Astrochelys radiata*). (A–D): Courtship sequences in repositioned from Black (1976, Figs. 1B and C). (D and F): Repositioned from Auffenberg (1978, Figs. 4A–C and 5). With permission from the *Great Basin Naturalist* and the Herpetologists' League, respectively.

populations. In a Macedonia population, for example, from 25% to 75% of female Hermann's Tortoises had cloacal injuries from the males' tail claws. The denser the population, the higher the injury rate, suggesting that the reproductive behaviors of Hermann's Tortoises evolved in lower-density populations.

Female Choice

Regardless of the males' approach to courtship, female tortoises behave in ways that prevent fertilization if they're not receptive. For the male to achieve copulation, the rear of the female's shell must be lifted off the ground. To frustrate a male's attempt at copulation, a female will press the rear of her shell against the ground by elevating her front, making it impossible for the male to insert his tail into her cloaca (see the Reproductive Organs section). It appears that female tortoises exert some degree of selection in choosing when to mate and with whom.

Also, in the wild, tortoises mate promiscuously with multiple partners during the reproductive season. In Mojave Desert Tortoises (*Gopherus agassizii*) and Hermann's Tortoises, for example, genotype analysis revealed that about half of clutches were sired by multiple males. Similarly, genotype analyses found evidence of multiple paternity in Steppe Tortoises (*Testudo horsfieldii*) in 27% of their clutches; in Gopher Tortoises in 28.6% of clutches; in Spur-thighed Tortoises in 20% of clutches. For female tortoises, polyandry (mating with multiple mates) may ensure more genetic variation in clutches, and thus increased chances of survival of at least some hatchlings under different environmental conditions.

Competition

Even though a male tortoise is not assured of sole paternity, males compete for females. Unlike female birds or mammals, female tortoises don't derive indirect benefits from mating events, like access to male territories or parental care (with a possible exception: males of certain species defending burrows or crevices). Generally, the female benefits are entirely from fertilization itself, that is, the genetic contribution of males to offspring. In this mating system, you would expect females to select males based on features that confer fitness, like hardiness. Even though male tortoises lack the showy plumage of birds, they may be advertising their quality as mates through the intensity of their courtship, such as the number of calls, as well as their success in male-male competitions.

Male tortoises demonstrate a variety of behaviors as they compete for females, ranging from head raising and bobbing to physical combat. Males of the six species of the North American group of *Gopherus* spp. tortoises have a natural weapon used in male-male combat. Called the gular horn, it's one extension that protrudes from the front of the plastron that is used to hook and potentially flip over a rival male (Fig. 4.4). Once on its back, a tortoise is vulnerable to predators and unsuitable tem-

Figure 4.4
Male Sonoran Desert Tortoises fighting for access to a female. The successful male has totally withdrawn his head as he prepares to flip the other male on his back. Sketch derived from a photo by Jerry L. Ferrara, Science Photo Library

peratures. Self-righting is essential but energetically costly for tortoise males that get overturned (Chap. 2).

Who Gets the Gal?

In many animals, larger males have an advantage in dominance contests. A study of Gopher Tortoises (*Gopherus polyphemus*) revealed that larger males fathered more offspring than smaller males. DNA analysis showed that one male individual had fertilized more than 70% of the eggs within a local population. The males that had sired eggs were significantly larger than other males, either because larger males prevailed in dominance contests or because females were exerting a preference for large males. Body mechanics may be another reason, as larger males can more easily get into a proper position for copulation. In Marginated Tortoises, mounting success of males increased the larger they were relative to the female. A smaller male is at higher risk of flipping over while attempting copulation (Chap. 2).

Experiments with Hermann's Tortoises have shown that males establish a dominance hierarchy, although it's not entirely based on body size. Alpha males achieve more copulation opportunities by being more aggressive toward other males as well as during their courtship of females. In some circumstances, however, tortoise females accept mates that are not alphas because they are superior in other health-related characteristics. In the Yellow-footed Tortoise (*Chelonoidis denticulatus*), for example, males engage in ritualized fighting. The winning males occupy better home ranges with more food and other resources. But these dominant males, because they're patrolling bigger areas, tend to pick up more parasites. Females will choose to mate with males that are lower on the dominance hierarchy but have lighter parasite loads. How a female detects the relative parasite loads of males is unknown, but could be from noting the male's vigor, odor, or some other salient feature.

Sexual Dimorphism

Natural selection for characteristics related to reproduction has led to sexual dimorphism in tortoises, that is, differences in physical characteristics of males and females. Typically, females are bigger than males within a species, perhaps because of selection for internal body space to contain eggs. In a population of Spur-thighed Tortoises in Algeria, 34 measurements of 67 individuals showed that females tend to have larger central scutes (from side to side), likely to accommodate eggs. In Hermann's Tortoise populations across their range—in Macedonia, Serbia, and Turkey—females are larger than males in nearly all body dimensions. Females also have heavier shells, perhaps providing protection for eggs, whereas males have lighter shells with larger openings that favor mobility.

If you look from a scute-by-scute perspective, however, even smaller-bodied male tortoises have some larger shell features that can be attributed to sexual selection, that is, selection related to competition for mates. Males tend to have larger front and rear scutes, which play a role in male-male combat and in copulation, respectively. Some tortoise species—such as *Gopherus* spp.—have an extension of their frontmost scutes called the "gular horn," which becomes more pronounced in mature males. This protruding structure just under the chin is used like a battering ram by males during combat.

Reproductive Organs, Copulation, and Fertilization

The Hardware

Like other vertebrates, male turtles have a paired reproductive tract with testes that produce sperm, which is stored in coiled ducts (epididymis) until mating occurs. Turtle testes, however, are located inside the body. A male turtle's tail is stouter and longer than a female's because it houses a penis versus a slightly smaller clitoris in females. Also, the male's vent is farther back on the tail than that of the female. The male must have a more distant vent to curve his tail under the female's and align his cloaca with her vent.

Tortoise females have paired ovaries, and like the male's testes, the ovaries lie at the top of the body wall on each side of dorsal vertebrae, in the rear third of the carapace and in front of the kidneys. Each oviduct lies immediately adjacent to its respective ovary. The oviduct enlarges with the development of the ovarian follicles. Each follicle encloses an unfertilized egg or ovum. Upon the maturation of the ova, they are shed by the ovary directly into the ostium (opening) of the adjacent oviduct.

Between the Sheets

A male tortoise extends his penis out of his vent and directly into the female's cloaca. The penis is long enough to reach into the urogenital sinus immediately adjacent to the openings of the oviducts. In an extreme case, the Aldabra Giant Tortoise (*Aldabrachelys gigantea*), the penis is over a foot long and bulbous at the end so it cannot be retracted until ejaculation. The female and male are locked together during copulation; pairs that fell sideways into coral crevices while attached have been found dead.

Once inside, the sperm moves upward in each oviduct to near the ostium (opening of duct to catch the ovulated ova). Ova must be fertilized at the top of the oviducts because as they begin their passage down the oviduct, each egg—assembly line style—receives a gelatinous coat prior to a fibrous matrix and calcareous coat that form the eggshell. The shelled egg(s) remain at the base of each oviduct until they are deposited through the cloaca into the nest.

Sperm Storage

Fertilization seldom occurs immediately. After mating, sperm is stored in folds in the upper part of a female's oviduct, from where they are released when she ovulates, or releases ova from her ovaries. Female tortoises of at least some species can store viable sperm for more than a year in these storage tubules and then still use it to fertilize eggs. Genotype analysis of Mojave Desert Tortoises revealed that sperm obtained in mating events two years earlier fertilized egg clutches. Sperm storage explains how females can lay clutches fertilized by the sperm from multiple males, not from a single mating event.

From a fitness standpoint, by not "putting all their eggs in one basket," females may be hedging their bets. Sperm from multiple males offers genetic diversity within her clutch, thereby increasing the potential that at least some of her offspring survive. It also confers the ability to synchronize egg laying and hatching with favorable environmental conditions by decoupling fertilization from mating. If the environment is stressful, female reproductive cycles may pause until conditions improve, such as in Radiated Tortoises (*Astrochelys radiata*), whose reproductive cycle is variable and only loosely associated with the timing of courtship.

Ovulation

On a broader scale, ovulation is seasonal and adapted to coincide with the optimal period for eggs to safely incubate and for hatchlings to encounter abundant food resources. Many tortoises ovulate in the spring after females emerge from hibernation, although there are variations, such as in the Sonoran Desert Tortoise (*Gopherus morafkai*), which ovulates once summer rains arrive. Ovulation requires advance preparation. In Mojave Desert Tortoises, the process of loading the ova with nutrients (vitellogenesis) begins in the fall and is completed in the spring after hibernation.

Female tortoise hormones—including estradiol, testosterone, and progesterone—cycle seasonally. So, females emerge from hibernation ready to mate. Increases in luteinizing and progesterone hormones spur ovulation. Because they need calcium to shell their eggs, circulating calcium also varies seasonally, as does vitellogenin, a protein molecule produced in the liver that is the major component of the egg's yolk in nearly all egg-laying vertebrates. Fertilized ova are enclosed in a layer of albumin before they move into the shell gland region of the oviduct where the eggshell forms, complete with a shell membrane beneath.

But even a female who has yolked her ova by spring will not necessarily lay them. If water and/or food are too scarce, female tortoises can conserve their resources by reabsorbing egg materials and foregoing laying.

Cycles: Prenuptial and Postnuptial

Tortoises have seasonal patterns of reproduction. Many have just one reproductive period per year, a "prenuptial" pattern. Gopher Tortoises, for example, mate only in the fall before hibernation, with females depositing a single clutch of eggs in the spring. Galápagos Tortoises (*Chelonoidis niger* spp.) do not hibernate, but they also have just one reproductive period. Male testosterone levels rise in both males and females and remain high through the mating period from August to November. During the nesting season of November through April, testosterone levels drop in both sexes, reducing sex drive.

Other species are "postnuptial," with two reproductive periods per year. For example, when Mojave Desert Tortoises emerge from hibernation in April, their testosterone levels are low and their testes have regressed, but sperm stored the previous fall in the epididymis are used in spring mating. During the summer, the male has no mature sperm, but his testosterone gradually rises in tandem with rising outdoor temperatures until he begins to produce sperm for the fall mating period as well as extra for use the following spring. In contrast, male Hermann's Tortoises in Yugoslavia begin sperm production and mating in April and May immediately after emerging from hibernation. Production continues through the summer, then peaks in August for the second reproductive period.

Nesting

The type of terrestrial habitat turtle females choose for nesting varies by species of turtle in relation to their ecosystems. Tortoises tend to live in arid or semiarid habitats, where they dig nests in well-drained, sandy soils. Gopher Tortoises in Mississippi seek out nesting sites near their burrows that have less vegetation, litter, and clay content in the soil. There are exceptions, however, such as tortoises that inhabit grasslands (e.g., Steppe Tortoise) or tropical rainforests (e.g., Red-footed Tortoise, *Chelonoidis carbonarius*); these species tend to nest in moist soils with more organic matter. A female must choose carefully, as the conditions of the nesting site are critical to the successful hatching of her eggs.

A Risky Business

The nesting process requires that females leave the safety of burrows or other refuges and risk exposure to predators or dehydration. Species that burrow (*Gopherus* spp.) may nest close to home, adjacent to their burrows or even in the entrances. Other tortoises travel long distances to find the right conditions, such as female Speke's Hinge-Back Tortoises (*Kinixys spekii*) that walk more than 400 m to nesting sites, almost twice the distance of their typical daily movements. If a female selects soil that is too dry or without the right proportions to make it structurally solid, the soil may give way and prevent a female from creating a properly shaped nesting chamber. Geometric Tortoises (*Psammobates geometricus*) will therefore travel farther to find suitable nest sites in the dry season as clay soils become hard and compact.

Tortoises often start digging a nest only to abandon the partially excavated hole before finding the proper soil conditions elsewhere to dig an egg chamber. Female Gopher Tortoises, for example, regularly make as many as three nesting attempts before digging an acceptable chamber. Female Speke's Hinge-Back Tortoises often move in small circles, seemingly to investigate a potential nesting site, then start digging it only to stop and move as far as 1.5 km to another site before successfully completing a nest and laying eggs. It's possible that these initial nesting attempts are hormonally priming a female for laying her eggs. Also, digging false nests may lure predators away from a real one.

Disturbances from weather, people, predators, or even other tortoises may also cause females to abandon nesting attempts. For example, time-lapse video documented a Gopher Tortoise in Georgia abandoning her nesting attempt during a heavy rainstorm. A tortoise may stop nesting when approached by a male tortoise or another female. Digging her nest is a dangerous and remarkable task for female tortoises.

Excavation Feat

Nesting is an amazing feat of excavation. We cannot know how the lizard-like precursors of turtles dug their egg chambers, although today's lizards dig nest holes with their front legs. Somewhere in the development of the plastron or carapace, the ancestors of turtles may have made the shift to digging nests with their rear legs. The nest is therefore dug sight unseen, using strictly tactile senses to shape the egg chamber into the flask-like cavity that expands outward at the bottom.

Most turtles have large, flexible hind feet with claws that can loosen the soil and gather it into the broad, scoop-like sole of the foot, then lift it out of the deepening chamber. Tortoises, however, must excavate using their elephantine hind legs with inflexible toes and shorter claws (Chap. 2), making the task even more laborious (Fig. 4.5). Each swipe of a hind foot removes only a small amount of soil, so many more digging strokes are required to create the egg chamber, occupying an hour or many more depending upon the crumbliness of the soil. Female Gopher Tortoises have been observed frothing at the mouth with exertion from preparing nests for laying eggs.

The digging is done alternatively: first one foot, then the other, until the cavity reaches the depth of the maximum reach of their hind legs. A tortoise does not stop but continues a lateral sweep of her feet at the bottom of the chamber to create a flask-shaped egg chamber. The tortoise's columnar foot precludes the broad lateral sweep, so her nest cavity remains largely cylindrical or lopsided cylindrical. The shape and size of a tortoise nest are variable depending on the interaction between a tortoise's body size and habitat. Species with longer legs dig deeper nests. Hard or compacted soils make digging more difficult and energetically more taxing, so a female may stop sooner. Berlandier's Tortoises (*Gopherus berlandieri*) are known to simply deposit an egg on the surface if the soil is too hard. If the soil is soft enough, some tortoises will dig a body pit that resembles a marine turtle nest.

Figure 4.5
Aldabra Giant Tortoise (*Aldabrachelys gigantea*) digging a nest with her hind legs after urinating, which makes the soil easier to remove and less likely to collapse into the egg chamber. Photograph by Ian Swingland

Burying the Goods

When finished excavating, the female positions her vent over the opening and expels her eggs into the chamber. With the eggs in the chamber, she uses her hind legs alternately to drag the excavated soil into the chamber (Plates 15 and 16). Once the chamber is filled, she tamps down the soil with her plastron and hind feet. The female Aldabra Tortoise urinates into the hole while digging to prevent the walls from collapsing and spreads the wet soil over a large area after nesting. Female tortoises often disguise the nest area by walking back and forth over it several times, perhaps to hide it from predators. Gopher Tortoises have been observed scraping the soil surface over a new nest with their nails, kicking soil from their burrows over it, and returning the same day or other days to continue to manicure the site. For at least some species, such as Aldabra Giant Tortoises, there is evidence that females return to the same nesting spot year after year, purportedly by smelling the urine left over from previous years. This "nest site fidelity" also occurs in other groups of marine and freshwater turtles.

The Primitive Way

The two *Manouria* species (Impressed Tortoise, *M. impressa*; Asian Giant Tortoise, *M. emys*) build nests not by digging but by making mounds of leaves and other surface debris. During the few days before egg laying, the female uses her front legs to sweep materials into a mound. Facing away from the mound, she sweeps the material along the side of her body, periodically moving backward until it reaches the mound. Observations of Impressed Tortoises in captivity show females clearing the area around their nests of leaves and debris as far as 10 m from the nest, like raking a yard. The female may continue some sweeping and shaping of the mound once she has laid the eggs. While tending the mounds, female *Manouria* tortoises also guard their nests from predators. Most females appear attentive for about a week, presumably until the odor of egg laying dissipates. Rarely does a female attend her nest for several weeks.

The Pancake Way

The Pancake Tortoise (*Malacochersus tornieri*) exhibits another odd way of digging a nest. A female often digs a shallow depression, just several inches deep, in loose, sandy dirt adjacent to her resident rock outcrop, or she may deposit her egg in a rocky crevice with no excavation at all. Female Pancake Tortoises usually produce just one egg per clutch, rarely two. But the crevice must have certain characteristics to be suitable—a deep, flat shape into which she can enter, whether horizontally or vertically. Pancake Tortoise crevices tend to taper to less than 5 cm in height, allowing them to wedge themselves in tight against predation (Chap. 5).

Laying the Eggs

A Tight Squeeze

In humans and other mammals, the pelvic girdle width limits the baby's head width, but the constriction is partially alleviated by softening the connection of the left and right halves of the pelvis. Turtles don't have that option because the left and right ilia are unmovable vertical posts; they're also anchored ventrally on each side by the pubis and ischium and dorsally by the sacral vertebrae and underside of the carapace (Chap. 2). To pass through the pelvic girdle, an egg diameter must therefore be less than the distance between the left and right ilia. The leathery shells of tortoise eggs may allow them to compress just a bit to pass through (Table 4.1).

There's one known exception, the Speckled Dwarf Tortoise (*Chersobius signatus*), a diminutive species just several inches long that lays a hard-shelled egg that's wider than its pelvic canal. Instead of the bony sutures of the pelvis, Speckled Dwarf Tortoise females have fibrous connections between the left and right halves of the pelvic plate that permit the stretching needed to pass an egg through. Increasing tortoise

Table 4.1.
Tortoise Eggs: Size and Shape in a Variety of Temperate and Tropical Species

Standard Name of Tortoise Species	Egg Size (width × length, mm)	Egg Shape	Female SCL (cm)	Country of Origin
Spur-thighed	28 × 34	near spherical	15.6–21.1	southern Spain
Hermann's	27 × 37	elongate	16.3–20.7	northern Greece
Asian Giant	?? × 44–58	oval	>50	Thailand
Travancore	40 × 50	elongate	25–29	India, captives
Burmese Star	40 × 55	elongate	>25	central Myanmar
Indian Star	27–39 × 36–52	elongate	25–32	India
Speckled Dwarf	~25 × 35	elongate	8.4–11.0	South Africa
Forest Hinge-back	31–34 × 34–39	elongate	~15–22 PL	Gabon
Pancake	22–39 × 37–50	elongate	12–18	Kenya and Tanzania
Angulate	28–38 × 39–45	elongate	>16.8	South Africa
Gopher	~43 × 43	spherical	23–25	Florida, USA
Chaco	~44 × 44	spherical	~13–21	Argentina

Note: The data derive from a single locality sampled by the researchers; thus the data may or may not be representative of the entire species. Egg shape is as stated by author of study. The width and length are given as the average egg size or as a range in a sample of several clutches. Abbreviations are as follows: PL, plastral length; SCL, straight line carapace length.

body size may relax this constraint on egg size by increasing the distance between the ilia, although the vertical distance between the carapace and plastron does not necessarily scale with body size.

That carapace-plastron distance may be (evolutionarily) shaped by other variables such as level of predation and can be potentially restrictive in some species. In these instances, the tortoises have evolved mechanisms to allow the passage of the egg between the rear (pygal) of the carapace and posterior end of the plastron (xiphiplastron). Hinge-backed Tortoises (*Kinixys* spp.) have developed the carapace hinge that allows the rear third of the carapace to tilt upward and widen the rear passageway (Fig. 4.6). Spur-thighed Tortoises have rear hinges along the plastral suture that allow up-and-down flexibility of the rear end of the shell. Female Berlandier's Tortoises develop a plastral hinge as egg laying approaches (Fig. 4.7). These hinges are functional only during egg laying season; the hormone relaxin temporarily softens the connective tissue sheath between the bones. In many species without flexible shells, females have a larger and/or deeper xiphiplastral notch, which increases the space into which the cloaca may stretch for the egg's exit.

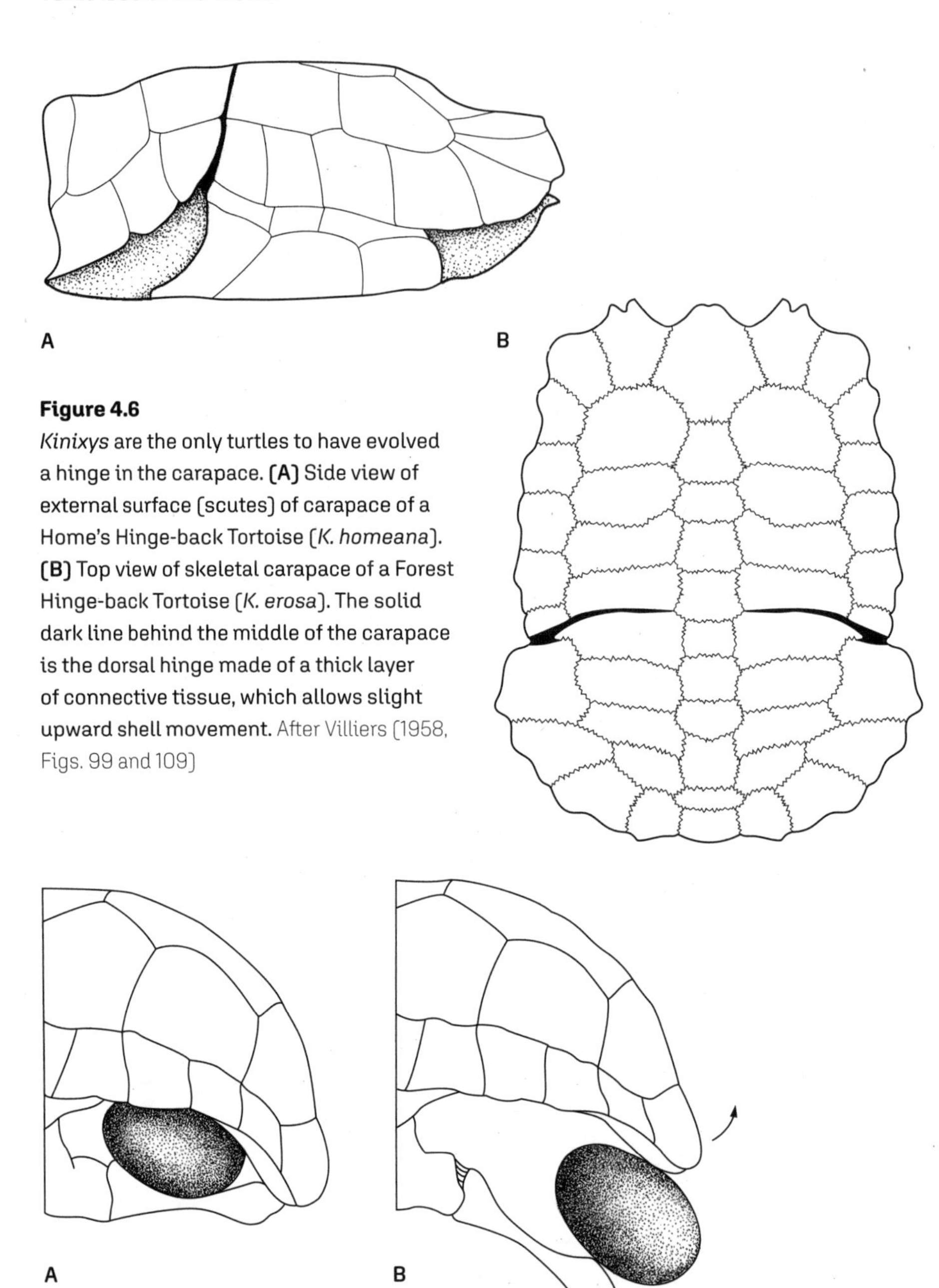

Figure 4.6
Kinixys are the only turtles to have evolved a hinge in the carapace. **(A)** Side view of external surface (scutes) of carapace of a Home's Hinge-back Tortoise (*K. homeana*). **(B)** Top view of skeletal carapace of a Forest Hinge-back Tortoise (*K. erosa*). The solid dark line behind the middle of the carapace is the dorsal hinge made of a thick layer of connective tissue, which allows slight upward shell movement. After Villiers (1958, Figs. 99 and 109)

Figure 4.7
In a few tortoises, the females develop a weakened suture between the third and fourth plastron plates (hypoplastra and xiphiplastral) as egg laying approaches. This weakening creates a hinge that allows the expulsion of the egg. The plastral hinge is especially well developed in Berlandier's Tortoises (*Gopherus berlandieri*), as shown here in a side view of **(A)** a female with egg inside and **(B)** a female laying an egg. After Rose and Judd (1991, Figs. 2A and B)

Tough Tortoise Eggs

As turtle eggs go, tortoise eggs are on the rigid end of the spectrum, that is, a hard-shelled egg with more calcium matrix. Eggshells have higher or lower calcium content depending on tortoise habitat. Whereas in birds, calcium occurs as calcite crystals, in tortoises, it's needlelike aragonite crystals. Aragonite plays an obvious role of protection, making the eggs harder to break. The variation in calcification of eggs may also reflect pH of the tortoise's nesting environment. Lower-pH (higher-acidity) environments require more calcified eggshells. It's possible that nesting soil acidity therefore influences clutch size because where less calcium is required per egg, larger clutches may be feasible. Tortoise eggshells are more waterproof than other turtle eggs, owing to a thicker calcareous layer that consists of more tightly packed crystals without pores, likely an adaptation to reduce water loss in their drier nesting habitats. Tortoise embryos rely mostly on the water supplied by the albumen and yolk, not on moisture from the soil around the eggs.

In just three known species of tortoises, the eggshell is bound by a cuticle, a layer on the shell surface. Cuticles generally isolate eggs from bacteria that could otherwise enter through the pores. Components of the cuticle—lipids, polysaccharides, proteins—have antibacterial properties. If you wash the delicate cuticle off chicken eggs, for example, they will lose their freshness quicker. Cuticles also play a role in gas and water exchange across eggshells. Notably, comparing cuticle thickness across the three tortoise species—tropical savannah Red-footed Tortoises (cuticle 26–20 μm), scrubland Indian Star Tortoises (*Geochelone elegans*; 32 μm), and dry forest Burmese Star Tortoises (*G. platynota*; 52 μm)—shows an increasing trend, probably tied to the risk of water loss.

Thickness of tortoise eggshells across species and habitats may also reflect a trade-off between calcifying eggs and laying multiple clutches. Eggshells from Speckled Dwarf Tortoises are surprisingly thin, lacking the multiple crystallite layers found in other arid environment tortoises. This species produces multiple clutches in a short spring nesting season. The benefits of multi-clutching might outweigh the benefits of investing the energy in increasing calcification.

Big Eggs or More Eggs?

Clutch Size

Although larger tortoises may not be able to lay bigger eggs, there is a body size advantage in other factors, including how frequently a tortoise lays and its total egg production per year. Tortoises generally lay many small clutches of eggs over a long period. Their reproductive span is longer than most other vertebrates since females remain fertile from maturity to death. Even though an individual clutch may fail, the high number of clutches over a female's lifetime ensures some successful reproduction. From an evolutionary perspective, their strategy suits the arid environments tortoises inhabit that go through whole seasons of drought.

Clutch size varies across species (Table 4.2). At one extreme, the smallest tortoise in the world—the Speckled Dwarf Tortoise (adults measure just 5–10 cm long)—lays just one egg in a nest. This single egg occupies up to 11.9% of her shell volume. At the other extreme is one of the larger species, the Asian Giant Tortoise (adults measure 46–60 cm long; Chap. 7), which lays 30–60 eggs per nest. Body size is clearly a factor in clutch size, but even the largest species, Galápagos Tortoises and Aldabra Giant Tortoises, lay clutches of no more than two dozen eggs. The Asian

Table 4.2.
Clutch Size in Various-Sized Species of Tortoises

Standard Name of Tortoise Species	Clutch Size	Adult Female Carapace Length (cm)	Country of Sample	Latitude of Sample
Spur-thighed	1–7	14.5–21.9	Spain	40°22′N
Hermann's	4–6	14.2–21.1	Greece	37°00′N
Burmese Star	1–11	>25	Myanmar	19°–22°N
Yellow-headed	1–9	24–33	Thailand	17°54′N
Asian Giant	23–45	53	Honolulu Zoo	captive
Radiated	1–5	24–36	Madagascar	25°36′S
African Spurred	13–31	34–58	Senegal	~15°N
Speckled Dwarf	1	8.4–11.0	South Africa	29°40′S
Home's Hinge-back	4–7	15–22	Nigeria	5°–6°N
Pancake	1	13.2–14.2	Kenya	2°–3°S
Leopard	3–7	38–78	Africa	~10°N–10°S
Sonoran Desert	1–12	22–29	Arizona, USA	32°–34°N
Gopher	4–12	25–33	Georgia, USA	31°59′N
Red-footed	8–13	21–44	Columbia	2°N–4°S
Chaco	1–7	15–33	Argentina	24°–34°S
West Santa Cruz	3–6	65–118	Galápagos Islands, Ecuador	0.6°S

Note: These data are from specific geographic populations of tortoises and are not a summary for the entire species. Clutch size is the number of eggs in single clutches laid by a sample of females from a single population. Many species lay more than one clutch of eggs in a single nesting season. The ranges of clutch sizes represent the variation of output for all females in a sample for a single nesting season.

Giant Tortoise may have the biggest clutch size because of the primitive characteristics of its genus (*Manouria*; Chap. 8), which also include cranial bone shape, lack of sexual dimorphism, wet forest habitat, and construction of aboveground nesting mounds.

Clutch Frequency

How frequently female tortoises lay is variable and responds to population density and quality of habitat. Females of some tortoise species lay multiple clutches of eggs about 30 days apart if environmental and nutritional conditions are favorable. During these rounds of nesting, ova are equipped with additional yolk. Studies of Aldabra Giant Tortoises found that as population density increased, both females and males began to reproduce later, and the number of clutches laid per year decreased. Males in high-density populations had smaller testicles, whereas at high densities or because of poor food availability after periods of low rainfall, females resorbed follicles, reducing reproductive rates. During periods of increased, unseasonal rain leading to more food availability, Aldabra Giant Tortoises laid more and larger eggs per nest. For a large, cold-blooded herbivore, tortoises are surprisingly well adapted to respond to changing conditions in their environments with respect to reproduction.

Habitat quality also plays a long-term, evolutionary role in the egg-laying patterns. North American tortoises that live in more predictable environments tend to reproduce less often, but with larger clutch sizes. Gopher Tortoises live in southeastern North America, where rainfall is seasonally ample and predictable. Thus the food supply is reliable, so they regularly produce a single, large clutch of five to eight eggs each year. In contrast, Mojave Desert Tortoises live in a dry region with pronounced seasons. Depending on rainfall, they produce up to three clutches per year of four to seven eggs each time. The Sonoran Desert Tortoise lives adjacent to the Mojave Desert Tortoise, but the Sonoran Desert habitat receives more predictable and abundant summer rains. Like Gopher Tortoises in the humid Southeast, Sonoran Desert Tortoises lay at most one clutch per year.

The thickness of tortoise eggshells may reflect a trade-off between calcifying eggs and laying multiple clutches under different habitat conditions. Speckled Dwarf Tortoises, which lay multiple clutches within their short spring nesting season, have surprisingly thin eggshells. They lack the multiple crystallite layers found in other arid environment tortoises. A short season implies a limited time for the female to garner scarce resources. So, for Speckled Dwarf Tortoises, the benefits of multi-clutching within one season might outweigh the benefits of better calcifying a smaller number of eggs.

Individual Variation

Even within a species, individual female tortoises vary widely in how many eggs they lay and with what frequency. No single factor can fully explain this individual variation, but certain factors stand out. First, larger females tend to lay larger eggs across diverse species such as Speckled Dwarf Tortoises, Hinge-back Tortoises, and Sonoran Desert Tortoises. The probability of laying eggs in a given nesting period also relates to body size; for example, larger Sonoran Desert Tortoise females are more likely to reproduce each year. But larger body size correlates poorly with the number of eggs in a clutch, which is also a determinant of a female's fecundity (reproductive success). How many eggs a female produces relates to her physiological condition, with healthier female tortoises laying more and larger eggs per clutch. In Sonoran Desert Tortoises, good maternal body condition makes for larger clutches of larger eggs.

Old Mamas

Female tortoises produce more offspring as they age. Their increasing reproductive prowess (fecundity) with age contrasts to that of mammals, whose fecundity declines over time. Compared to younger females, older Spur-thighed Tortoise females have more offspring with higher survival rates in their first season of life (i.e., from hatching to overwintering). An older females' offspring may be smaller bodied upon hatching (thus making a larger clutch size possible) but may catch up in size during their first year.

Egg Mortality

Multiple factors affect tortoise egg mortality. Eggs are occasionally broken by the female as she nests, but it is uncommon. Of 156 Spur-thighed Tortoise eggs monitored in southwestern Spain, only 6 eggs (3.8%) were broken during laying. Most egg mortality is from predators or unsuitable nest conditions.

Egg Predators

Tortoise nests are highly vulnerable to predation by a range of animals that dig, such as raccoons, badgers, or wild hogs. Asian Forest Tortoises in India commonly lose nests to water monitors and nocturnal civets. The effects of predation vary. A study of Spur-thighed Tortoises found that only 4.5% of eggs were lost to nest predators, while nest predation eliminated more than 50% of eggs in Gopher Tortoises, Mojave Desert Tortoises, and Galápagos Tortoises. For Yellow-headed Tortoises (*Indote-*

studo elongata), hatching success was 81% in the absence of predators but reduced to 60% by predation. In some settings, nest predation is such a significant source of mortality that it's implicated in population declines. In one such case, badgers destroyed 95% of Hermann's Tortoise nests in a locality in southern France.

High nest predation rates may result from the complex and historical interplay of tortoise habitat and human uses. For example, suitable nesting habitat for Hermann's Tortoises in southern France became scarce, in part because humans have stopped clearing areas for horticulture. The short-term result was an increased nesting density and consequently higher vulnerability. On Aldabra, the main predator is the Coconut Crab (*Birgus latro*), the world's largest terrestrial crab. Coconut Crabs are attracted to laying females (Plate 17) and not always confused by the spreading of soil to hide the nest. Historically, when Aldabra Giant Tortoises were abundant, pigs introduced from sailing ships feasted on their eggs as they dropped from the females' tails.

Even in lower-density situations, tortoise eggs are vulnerable simply because of their high nutritional value to predators. The mob of predators that eat Gopher Tortoise eggs include bobcats, raccoons, Virginia opossums, foxes, coyotes, armadillos, and skunks. Experiments in which exclusion fences were used to prevent egg predation resulted in increases in hatching success from 37.5% to 74.4%. Predators, especially native species, may track the seasonal rhythms of tortoises to target them during nesting periods.

Nest Conditions

In the absence of predation, hatching success of tortoise eggs still varies according to ambient conditions. Female tortoises select nest sites based on sun exposure, slope, and variables like soil composition and moisture that are ascertained by sniffing. Still, female tortoises are not infallible in choosing nest sites. Subtle factors that cannot be anticipated also affect hatching success, such as parasite attacks or changes in weather. Gopher Tortoise eggs have a demonstrated success rate in natural nests of 16.7%, but in laboratory experiments where ambient variables were controlled, hatching success averaged 58.8%.

Nest Guarding

Tortoises (and turtles generally) don't typically protect their eggs. Although it is rare, turtles in a few families—including tortoises—have female nest care. An isolated report appears in a 1923 paper of a female Leopard Tortoise (*Stigmochelys pardalis*) staying near her eggs. Several species of *Gopherus* regularly exhibit nest guarding. Mojave Desert Tortoise females build their nests close to their burrows

(Chap. 6). As they excavate a burrow, the sand tossed aside piles up outside the burrow entrance into a mound or "apron." Females sometimes use the apron as a nesting site. In one instance, researchers observed a female tortoise apparently defending her eggs by hissing, lunging, and attempting to bite biologists as they excavated an apron nest. In other observations, a female Mojave Desert Tortoise stretched her legs out, wedged herself on top of the nest, and tamped more dirt down into the burrow. Female Sonoran Desert Tortoises have been observed aggressively trying to repel Gila monsters approaching their nests as they lay eggs. Gopher Tortoises have also been observed nest guarding.

The most protective female is the Asian Giant Tortoise, who physically guards her nest for at least a few days after laying. She shoves approaching predators with the front of her shell. If the threat gets too close, the female lays on top of the nest with her legs splayed and refills any holes made by predators with more nesting soil and leafy debris. After a few days, as the odor from egg laying abates and as the nest becomes increasing camouflaged with newly accumulated leaf fall, she stops guarding it. One "helicopter mom" was recorded guarding her nest for six weeks after laying her eggs.

Embryo Development and Hatching

As a turtle embryo develops inside the shell, its yolk sac serves as a source of energy and gradually shrinks. The ossification of a turtle bones happens quickly, beginning first in the skull and jaw bones, followed by the plastron (bottom shell). When the hatchlings emerge from their eggs, the main skull bones, leg skeleton, and plastron are partly ossified. The carapace is less ossified at hatching.

Waiting to Hatch

The time from egg laying to hatching—the incubation period—varies widely in response to various factors, including species, specific maternal genetics, and environmental conditions including nest temperature. Incubation times of wild Spur-thighed Tortoises in southwestern Spain ranged from 78 to 118 days, with eggs laid earlier in the season incubating longer, likely because of exposure to lower temperatures at the outset. Red-footed Tortoise eggs hatched in captivity show that incubation temperatures influence early growth rates; eggs at 29.5°C hatched significantly more quickly (141 days) than eggs at 27.5°C (201 days). Higher-temperature incubation yielded a lower hatching rate (only 64% of eggs hatched versus 100% at lower temperature), although the hatchlings were larger and heavier at three months of age. So, a higher, but still suitable, incubation temperature results in fewer, larger hatchlings.

The time to hatching varies depending on nesting conditions. In Ploughshare Tortoises (*Astrochelys yniphora*), eggs incubate for as long as 281 days, with eggs deposited earlier in the season having the longest incubation. Hatchlings emerged at the onset of the rainy season, suggesting that water availability stimulates hatching. Development in these situations of long incubations is not continuous but reaches a mid-embryonic stage and pauses. This halting of development—embryonic diapause—occurs in other species besides the Ploughshare Tortoise (Chap. 5).

Hatchling Sex Ratio

Incubation temperature also controls the sex of hatchling tortoises in the species that have been studied to date. Marine turtles and many freshwater turtles have temperature-dependent sex determination, where males typically result from lower incubation temperatures and females from higher temperatures. Nine tortoise species are known to have temperature-dependent sex determination: Pancake Tortoises, Mojave Desert Tortoises, Gopher Tortoises, Spur-thighed Tortoises, Hermann's Tortoises, African Spurred Tortoises (*Centrochelys sulcata*), Aldabra and Galápagos Giant Tortoises, and Red-footed Tortoises.

In Mojave Desert Tortoises, for example, incubation temperatures near 30.5°C or below produce all males, whereas incubation temperatures of 32.5°C or above produce all females, with a mixed sex ratio for the middle degrees. But the male-female temperature thresholds vary across species, depending on temperatures in their natural environments. For Red-footed Tortoises (endemic to cooler forests; Chap. 7), the hatchlings were all males when the nest was incubated below 24.34°C and all females at temperatures above 27.9°C. The feminizing temperature of 32.5°C for Mojave Desert Tortoises would be lethal to the Red-footed Tortoises.

A study of African Spurred Tortoises tested several post-hatching physiological parameters and found no difference relative to incubation temperature. On that basis, the researchers hypothesized that temperature-dependent sex determination is retained because natural selection against it is either weak or nonexistent, not because it confers an advantage.

A Reproductive Oddity

Most turtle embryos, including those of tortoises, begin cell division while the fertilized ova (zygotes) are in the females' oviducts; however, cell division stops at the gastrula stage, an early embryonic stage well before tissue differentiation and organ formation. The pause is related to low-oxygen (hypoxic) conditions. Development resumes only when oxygen becomes available again, typically when the eggs are deposited in the nest.

There is one tortoise species known to bypass the arrested development at the gastrula stage. The Angulate Tortoise (*Chersina angulata*) can retain eggs in her body before laying for exceptional long periods lasting from as little as 23 days to as many as 212 days. At the longer end of the range, female Angulate Tortoises retain eggs right up until they're ready to hatch. For example, an Angulate Tortoise in a captive colony in South Africa held on to her last clutch until a few days before the eggs hatched. During an unusually hot summer, rather than depositing her eggs in a nest, she incubated them in her body. This "facultative viviparity" in response to heat might have protected the hatchlings from excess heat exposure. An Angulate Tortoise can exert some control over egg incubation temperature by keeping the eggs in her body and shuttling from sun to shade.

This adaptation has not been recorded for any other tortoise, or indeed any other turtle. It is another example of the evolutionary adaptability of turtles and offers insight into how turtles have been able to persist for millions of years through some extremely harsh climatic events.

CHAPTER 5

Hatchlings to Adults

Tortoise Life Cycle

Tortoises are famous for their longevity, even if some species are not long lived. That said, tortoises face many risks, and most individuals don't make it to adulthood. They are vulnerable as barely developed embryos within eggshells that are in a hole in the ground until they reach a carapace length of 20–25 cm that's more predator resistant (Plate 1). Their notorious longevity derives from the few captive individuals that have lived for more than a century. One famous Spur-thighed Tortoise (*Testudo graeca*)—thought to have been captured in about 1844 from the Turkish shores—spent a half century aboard British Navy ships, then another century in the rose garden of Powderham Castle in Devon. "Timothy" was about 160 years old when she died. A Galápagos Giant Tortoise (*Chelonoidis niger*) that was brought back to England by Charles Darwin ended up at the zoo in Brisbane, Australia. Known as "Harriet," she was estimated to be 176 years old when she died. The record-holding Aldabra Giant Tortoise (*Aldabrachelys gigantea*), "Jonathan," just celebrated his 190th birthday on St. Helena, where, already about 50 years old, he was given to the governor in 1882. These centenarian tortoises are remarkable, but even more remarkable is the ability of half-century-old wild tortoises to survive environmental hazards and continue to reproduce for decades more. Tortoise life histories, although variable, share the characteristic of reproduction into old age.

Inside the Egg

Nourishing Yolk

Both inside and outside of the eggshell, the embryonic and hatchling tortoise is nourished by yolk deposited in the ovum (egg cell) before fertilization. The embryo will complete its entire development within the confines of the eggshell, relying on the packed nutrients (platelets) from the yolk. While in the oviducts, the low-oxygen environment apparently arrests the embryo's development. Once the egg settles in the bottom of the nest chamber, both the stationary position and higher oxygen content of the nest trigger development. Some of the yolk is metabolized to produce embryonic tissues. Other yolk that is not metabolized becomes part of the embryo's developing digestive tract (yolk sac) and will serve as the main source of nourishment for the hatchling once it leaves the egg. This nutrient supply ensures the hatchling has a food supply until it can feed on its own.

Looking Like a Tortoise

The sequence of development is similar in all reptiles, including birds. It proceeds from a ball of embryonic cells (the gastrula) that differentiate to take on specific tissue functions and multiply in number. The gastrula begins to elongate and forms a neural tube that will become the central nervous system. Shortly thereafter, leg buds emerge, fore and aft. As the external morphology is developing, so are the internal organs. Step by step, organ by organ, the undifferentiated ball of cells becomes a recognizable tortoise, with its carapace and plastron formed but largely unossified (not yet bony). Inside the rigid and spherical eggshell, the embryonic tortoise folds in on itself with its snout close to or touching its tail.

Toward the end of development in the egg, turtle embryos shift from the usual reptilian pattern. The embryonic body walls grow outward rather than downward and encompass the pectoral girdle inside of the ribs (a peculiar arrangement unique to turtles). Another unique bulge of cells (carapacial ridge) forms along the fold line on each side, which then drives the further development of the shell. The ribs change their growth pattern to produce a domed shell shape rather than a cylindrical trunk. This developmental shift is considered a key step toward the distinctive turtle body plan.

Getting Protective Scutes

The scutes develop from outer dermal tissue (ectoderm; Chap. 2). A species-specific scute design that is genetically coded gives each species a different shell color pattern, such as the yellow markings on the Indian Star Tortoise (*Geochelone elegans*). The scutes start as thickenings of epithelial cells (placodes) that are also the basis of the scales of lizards, crocodiles, and birds. Just a bit after the carapacial ridge shows up in turtle development, the scute placodes form, first for the carapace and then for the plastron. In some turtles, mostly soft shells but also in Spur-thighed Tortoises, small, hard bumps (tubercles) develop on the carapace from additional thickenings of the top layer of epithelial cells. In tortoises, the bumps are small, as they are contained by a strong keratin layer in the epidermis.

Hardening Up the Shell

Still, a hatchling turtle has not completed the development of its carapace. Upon hatching, the shell is soft and will continue ossifying (hardening into bone) for at least the first year of life. In tortoises, ossification takes longer. Studies of the ossification of Hermann's Tortoises (*Testudo hermanii*) showed that ossification continually increases for the first eight years of life. Growth of the bony costal and neural plates starts at birth, but the fontanels (spaces between the bony plates) do not disappear altogether until nine years of age. Ossification rates vary depending on tortoise species, and a similar ossification pattern has been seen in fossil species, such as *Changmachelys bohlini* from the Early Cretaceous.

Some unusual variations in the development of the shell relate to species-specific adaptations. For example, Africa's Hinge-back Tortoises (*Kinixys* spp.) do not have hinges at hatching. The hinge is absent in juveniles and arises through a rearrangement of bony and keratinous tissues as it grows. Pancake Tortoise (*Malacocherus tornieri*) shells remain flexible throughout their lives, an adaptation to taking refuge in rock crevices. When scientists first found them, they assumed that the reduced ossification was a genetic or pathological defect, but the limited ossification of their flat shells allows Pancake Tortoises to compress when they wedge into crevices to evade predators. Their shell ossification is arrested partway through development, leaving parts of the shell cartilaginous and somewhat transparent.

Incubation

The normal incubation period of clutches for any tortoise species varies. For the European *Testudo*, incubation varies from ~60 to 130 days, for example, 67–129 days in the Spur-thighed Tortoise and 58–100 days in Hermann's Tortoise. The North American Gopher Tortoise (*Gopherus polyphemus*) incubates for 97–106 days, and its cousin the Mojave Desert Tortoise (*Gopherus agassizii*) incubates for 67–104 days. Studies of Yellow-headed Tortoises (*Indotestudo elongata*) at different sites yield incubation times of 42–56 days, 98–150 days, and 117–180 days. But the African tortoises that have the most variable and lengthy incubation periods are 72–116 days for the African Spurred Tortoise (*Centrochelys sulcata*), 105–129 days for the Speckled Dwarf Tortoise (*Chersobius signatus*), 197–281 days for the Ploughshare Tortoise (*Astrochelys yniphora*), and a whopping 247–324 days for the Spider Tortoise (*Pyxis arachnoides*). This high variability shows that there is no standard incubation period for tortoises, not even within a species. Why so much variation? The climate of their habitats and their adaptations to ambient conditions appear to shape the duration of incubation.

Triggers for Hatching

Hatching is commonly synchronized with seasonal environmental conditions. For many tortoises, hatching is triggered by the beginning of the wet season, African Spurred Tortoises, Leopard Tortoises (*Stigmochelys pardalis*), and Radiated Tortoises (*Astrochelys radiata*), for example. This timing is essential to ensure that hatchlings emerge from the nest when there is abundant plant growth, especially of young, succulent seedlings. In some especially dry settings, both the adult reproductive season and hatchling emergence coincide with the wet season (typically hatching at the beginning and egg laying at the end). The result is an exceptionally long incubation period that lasts through the dry season. Spider Tortoises and Flat-tailed Tortoises (*Pyxis planicauda*) in Madagascar, for example, have a long incubation period that corresponds to the lengthy dry season.

Taking a Break

The key adaptation for this extended incubation may be the phenomenon of developmental arrest or diapause, which is often associated with insects temporarily halting development in cold-temperature environments. For tortoises, long incubation periods are typically associated with arid environments and likely result from developmental arrest, although such arrest has not been proven yet for tortoises. Similarly, we do not know what factors stop development and restart development, nor at what development phase it begins, just that a shift back to warmer, humid conditions appears to trigger the completion of embryonic development sooner. Laboratory evidence shows that cooling the eggs of Spider Tortoises for six to eight weeks during development results in successful hatching, suggesting they may undergo diapause in the wild. But for such an important feature of many tortoises' life histories, we know surprisingly little, other than it occurs for tortoise species with half a year or more between egg deposition to hatchling emergence.

Hatching

For turtles, hatching and emergence from the nest occur at separate times. Unlike a hatched chicken, which is immediately active and ready to feed, a hatched tortoise may spend days, weeks, or months underground in the nest before emerging. Indeed, the time between a hatchling first cutting open its egg and getting out of the eggshell is often prolonged. In Angulate Tortoises (*Chersina angulata*), eggshells were observed cracking more than two months (67 days) before hatchlings exited.

Getting Out of the Egg

The mechanics of turtles getting out of their eggs helps explain why it can be a lengthy process. When a tortoise is ready to hatch, it uses a sharp keratinous addition at the tip of its beak to slice or puncture the shell (Fig. 5.1). Although often and incorrectly called an "egg tooth," it's not a dental structure but rather a temporary sharp keratinous bump, "the egg carbuncle." As turtle embryos develop, the embryos extract calcium from the eggshell, thereby thinning the shell and making it weaker and easier to break through at hatching. A combination of "pipping" with the carbuncle and pushing with its legs over a period of hours or days frees a hatchling tortoise from its shell. The carbuncle eventually wears away.

Tortoise Shell Architecture

Not all tortoise eggshells are alike; rather, they vary with incubation environments. Typically, dryer or lower-pH environments warrant denser eggshells to protect the embryo; however, there are exceptions. The eggshells of Speckled Dwarf Tortoises are surprisingly thin, considering that they inhabit the hot, dry Namaqualand of South Africa. The reduced calcareous layers may reflect the brief inter-nesting periods for females. Females lay multiple clutches within their short spring breeding season; this likely does not allow much time for extensive calcification. Thinner eggshells are a trade-off with producing more eggs.

Figure 5.1
A hatchling Galápagos Tortoise (*Chelonoidis niger*) emerging (hatching) from its eggshell. Note the pointed carbuncle on the tip of its snout used to slice the shell. Sketch derived from www.GiantTortoise.org

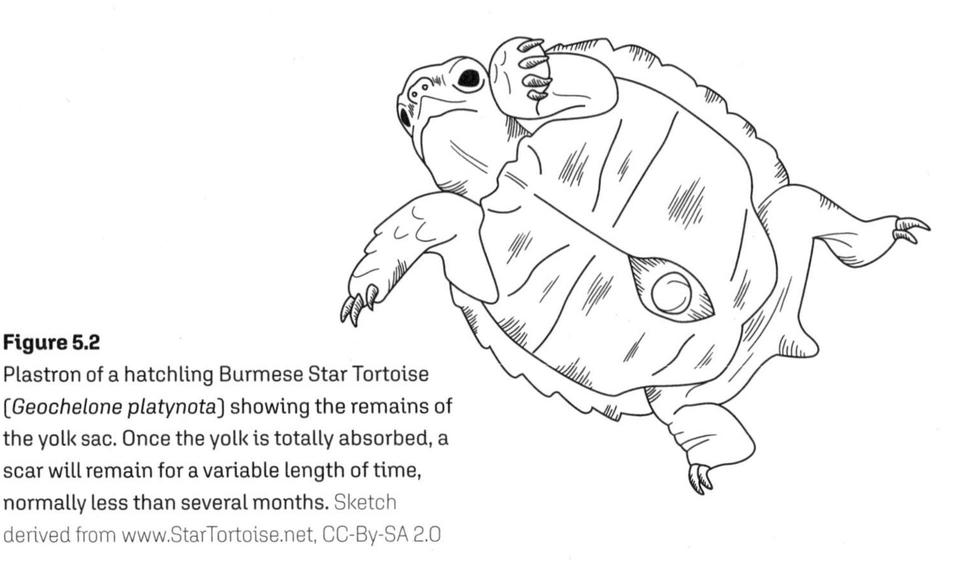

Figure 5.2
Plastron of a hatchling Burmese Star Tortoise (*Geochelone platynota*) showing the remains of the yolk sac. Once the yolk is totally absorbed, a scar will remain for a variable length of time, normally less than several months. Sketch derived from www.StarTortoise.net, CC-By-SA 2.0

Upon hatching, the yolk sac is too large to be contained within the body cavity, so it protrudes outside the plastron through an opening (the umbilicus) (Fig. 5.2) As its nutrients are extracted, the sac shrinks, but at hatching, part of the sac visibly pokes out of the plastron of most tortoises. Even though the external portion of the yolk sac limits movement of hatchlings by making walking difficult, it's an essential nutrient source. Typically, the yolk sac is withdrawn into the body within a few days, and the umbilicus opening closes.

Hatching Success

The proportion of tortoise eggs that hatch from a given clutch of eggs is highly variable, even ignoring losses to egg predators. For example, a study of Gopher Tortoises recorded a 90% hatch rate one year and only 50% the following year. Estimates of hatching success for Mojave Desert Tortoises range from 46% to 83% across geographic sites.

Hatching success likely correlates with a suite of factors, including female age and health as well as nest conditions. In a population of Yellow-headed Tortoises, the hatching rate for nests that were not disturbed was 81% but dropped to 60% when nests suffered disturbances from people or chickens. Wild hatchling turtles often have shell abnormalities (frequency ranges from 2% to 69%). Many are anomalies in scute shape and number and do not appear to be life threatening. Some, such as a cleft palate or being born with two heads, affect fitness. The abnormalities may stem from incubation conditions such as humidity or presence of infectious pathogens. When selecting nest sites, females cannot predict the ambient conditions that will prevail for the duration of incubation (Chap. 4).

Getting Out of the Nest

Tortoise hatchlings often spend some time in the nest before making their way to the surface. Hatchlings may delay their emergence until they sense better environmental conditions for survival. The variable lengths of time to emergence, even within a species, suggest that hatchlings have some flexibility to emerge when ambient conditions are favorable. Spur-thighed Tortoises in Spain, for example, emerged anywhere from 1 to 23 days after hatching. A Galápagos Giant Tortoise hatchling spends up to a month in the nest until all the eggs have hatched, and then all siblings emerge simultaneously. Hatchling Ploughshare Tortoises from the same clutch were also reported emerging within a single day or at most two days, whereas emergence of a clutch of Yellow-headed Tortoises occurred over a 12-day period.

This plasticity in hatching times and emergence from the nest is undoubtedly a tortoise adaptation for responding to unpredictable environmental conditions. Emergence is not always successful, though. For example, unusually high rainfall caused almost 10% of Mojave Desert Tortoises hatchlings to become entombed in the nest (drowned or unable to exit through the weight of the waterlogged sediment). There is some evidence that larger hatchlings have an advantage in exiting from nests. Thus larger (older) females laying larger eggs or producing larger hatchlings may have higher hatchling survival rates.

First Days

Independence

Hatchling tortoises are on their own once they leave the nest. Turtles are not known to do any parental care other than egg guarding in some species (Chap. 4), so hatchlings must be fully self-sufficient. For a tortoise, that typically means foraging for plants, mushrooms, and invertebrates, as well as regulating its body temperature by seeking suitable amounts of sun or shade. It's remarkable that a newly hatched tortoise can distinguish between what's edible and not edible. It may be cueing on color. Experiments with Leopard Tortoises showed their preference for red, light green, and olive. Young tortoises also likely use their sense of smell, as evidenced by hatchlings first poking a potential food with their noses before biting. Finally, there are likely elements of trial and error.

Newly emerged hatchlings tend to quickly seek refuge, although they must also spend a substantial amount of time foraging to support their fast growth. There's limited evidence for dispersal of hatchling tortoises from the nests, perhaps in part because of the challenge of observing this small, cryptic life stage. Gopher Tortoise hatchlings typically move a few meters from the nest and then excavate their own, hatchling-sized burrows—small, hidden openings concealed under vegetation, selecting areas where there is lots of ground cover. Like adult gopher tortoises, juveniles and even hatchlings will compete for burrow space. Recordings inside burrows showed young Gopher Tortoises shoving, biting, and flipping each other over as they contest for burrow space.

Daily Habits

Hatchling tortoises quickly adopt a daily pattern of activity like that of the local adult tortoises. Hatchling Mojave Desert Tortoises, for example, come out to forage midday, during the warmest time in spring, and take refuge in late afternoon through the night. When temperatures heat up for summer, however, they switch to a bimodal pattern of foraging in the morning and early evening to avoid the too-hot middle part of the day. In contrast to adult Mojave Desert Tortoises, some juveniles of this species are occasionally active during the winter. Perhaps their smaller body size allows them to heat up more quickly on unseasonably warm days.

There's a smattering of evidence that tortoise hatchlings can recognize familiar tortoises (kin recognition). In an experimental setting, hatchlings of *Testudo* species—Marginated Tortoises (*T. marginata*) and Spur-thighed Tortoises—sniffed, then avoided unfamiliar hatchlings of the same species. In contrast, they totally ignored familiar hatchlings. Kin recognition in turtles, in general, has been detected in studies over the past decade, and it's reasonable to predict that kin recognition occurs between tortoises of various ages based on smell.

The Odd One

Most aspects of the Pancake Tortoise are unusual, including its hatching and first days of life outside the egg. Females lay their eggs either in loose soil or within a rock crevice. Females typically lay just one egg per clutch but nest multiple times during the four- to six-month season. The embryo incubates in the egg for several months up to a whopping 237 days. Her lone hatchling emerges tightly curled but flattens out within a day.

Hatchling Survival

Whether hatched in a crevice or underground, hatchlings are preyed on by a wide range of other animals. Adult tortoises suffer predation, but hatchlings and juveniles are much more vulnerable. Both native and introduced predators are also a significant source of mortality of tortoise eggs and hatchlings. Galápagos Tortoise nests are predated by pigs, dogs, cats, and black rats that were brought to the islands aboard ships.

The Hazards Start Early

The vulnerability of tortoises begins when they're still in the egg (Plate 17). Once a turtle has pipped a hole in its egg, chemical odors may attract predators. Larger animals such as raccoons will dig for eggs and hatchlings. Wild African Bush Pigs (*Potamochoerus porcus*) in Madagascar have devasted nests of Angulate Tortoises. But reptiles, raptors, and even insects are a problem, too. Two-banded Monitor Lizards (*Varanus salvator*) swallow whole Asian Giant Tortoise (*Manouria emys*) eggs. Introduced Red Fire Ants (*Solenopsis invicta*) may be responsible for more than a quarter of the mortality of Gopher Tortoise hatchlings. Besides swarming over hatchlings and eating them from the outside in, fire ants may get into nests and damage eggs. Even if hatchlings survive a swarm, they may have reduced growth rates.

Newly emerged hatchlings often face a suite of predators (Plates 1 and 18). Hatchling Egyptian Tortoises (*Testudo kleinmanni*) are prey for birds of prey, monitor lizards (*Varanus* sp.), hyenas, and wolves. Hatchling Angulate Tortoises face danger from predators native to their southern African habitats, such as baboons (*Papio* sp.), rock monitor lizards, mongooses, jackals, and Secretary Birds (Sagittarius serpentarius). A radiotracking study of Gopher Tortoises in southern Mississippi found that most hatchlings (65%) had been killed by predators within 30 days of hatching. The deaths were mostly attributable to mammals, although some were from fire ant attacks. Only one hatchling of 48 made it to day 736 (approximately two years old). Another study that tracked 85 Gopher Tortoises hatchlings in central Florida reported that not a single tracked hatchling lived more than 335 days. Most were killed by mammal and snake predators.

Birds in the *Corvus* genus are increasingly widespread and particularly adept at preying on hatchling tortoises of various species. Common Ravens (*Corvus corax*) are responsible for an estimated 70%–91% of the mortality of Mojave Desert Tortoise hatchlings. A study of Spur-thighed Tortoises in Morocco showed Common Ravens to be the cause of all observed mortality of hatchlings, with deaths lower only where hatchlings could hide under vegetation. Ravens hunt by sight, sometimes in pairs. They spot and then stab prey with their powerful beaks. Similarly, through predation on hatchlings, their cousins the Pied Crows (*C. alba*) in South Africa are implicated in population declines of Speckled Dwarf Tortoises, while White-necked Ravens (*C. albicollis*) may be actors in the demise of juvenile and adult Karoo Dwarf Tortoises (*Chersobius boulengeri*) (Chap. 9).

Bigger Is Better for Survival

Studies show a positive correlation between hatchling size and survival rate during the first year: bigger hatchlings have better chances of making it to their first birthday. In Mojave Desert Tortoises, every millimeter increase in the mean carapace length of hatchlings doubles their chances of surviving the first year. Larger hatchlings may have advantages in getting food resources, storing water in their bladders (Chap. 3), and/or escaping predators. Or they may exit their eggs already provisioned with more energy reserves. Given that hatchling size correlates with egg size, which correlates with the mother's body size, you'd expect selection for larger female tortoises. Indeed, in many tortoise species adult females are larger than adult males.

Hatchlings, like adults, also perish from human-related hazards, such as deaths of Yellow-headed Tortoises that get hit by cars or trampled by large cattle in a village in northeastern Thailand. Of 100 tracked Yellow-headed Tortoise hatchlings, a third died during a three-month period, the majority from unknown causes, possibly parasites or excessive heat or dryness. The bottom line is that being a tortoise hatchling is hazardous. High hatchling mortality is a characteristic of tortoises, even in stable populations. The challenge is detecting when introduced predators or increased exposure to native predators causes tipping points in populations from which they cannot recover (Chap. 9).

Growth

Growth of tortoises is fast when they are young and slows down as they age. Not surprisingly, we measure growth by the increasing length of the carapace, and occasionally by body weight, but the latter is less reliable than carapace length. The normal pattern in all vertebrates is a rapid increase in younger animals, with growth slowing as individuals approach sexual maturity, then stopping or becoming very slow. As carapace length increases, growth rate decreases, although male tortoises may have growth spurts as they approach sexual maturity. For example, a study of Gopher Tortoises revealed a growth rate of 24.2 mm/year for hatchlings, 9.3 mm/year for juveniles, and 1.6 mm/year for adults. Adults continue to grow, but at increasingly slower rates, because more energy is allocated to reproduction and physical maintenance.

Natural Variability

For tortoises, growth rates are also linked to environmental factors like temperature, rainfall, and abundance of forage. In the semiarid and arid climates inhabited by most tortoises, growth is slow, which also means it takes a long time to reach maturity. Female Speckled Dwarf Tortoises in the arid Karoo of South Africa, for example, take 11 years to reach maturity. Available nutrition affects growth rates, based on captive husbandry data, such as with Red-footed Tortoises (*C. carbonarius*) that grow faster but with lower bone density on a diet high in starch and low in fiber.

Spur-thighed Tortoises in Morocco had variable growth patterns across habitats. At three sites, growth was relatively fast for the first 7–10 years after hatching and then decreased rapidly. But within that broader trend, females and males grew at the same rate at one site, in contrast to the other sites where males grew more quickly than females. Males reached sexual maturity at 6–7 years old at a smaller size, compared to females who matured at 9–10 years old and at a larger size. Clearly, environmental conditions in the wild affect food availability and therefore tortoise growth.

Pauses in Growth

Growth may halt altogether under drought conditions. During an especially dry year with minimal rainfall, some Speckled Tortoises showed negative growth resulting from internal tissues shrinking from dehydration and starvation. A period of poor nutrition may even cause bone resorption, thus reducing the size of a tortoise. The shrinkage is reversible, with Mojave Desert Tortoises showing carapace length shrinkage during a drought year, followed by restoration of original size during a rainy year as its tissues gradually rehydrate and bounce back.

Because of tortoises' resilient physiology, they can weather the extremes of a semiarid habitat. As climate change brings more intense or prolonged dry spells, however, even drought-tolerant species like Speckled Tortoises will feel the effect (Chap. 9). Females that have shrunken in response to drought produce smaller eggs and thus smaller hatchlings that are more vulnerable.

Sexual Maturity

Tortoise growth rates go hand in hand with attaining sexual maturity. Roughly, female turtles mature when they've reached about 70% of their maximum body size. So, tortoise females, regardless of their species' size, have a similar value (~70%) for body size at maturity relative to their potential maximum size. But in some settings, growth rate may overshadow body size in signaling sexual maturity. In a population of Steppe Tortoises (*Testudo horsfieldi*), precocious individuals that grew the fastest became sexually mature at a smaller size than slower-growing individuals. Often the smallest tortoise female in a population is the oldest, having required more years at a slower growth rate to reach maturity.

The age of sexual maturity is variable depending on species. In a population of

Berlandier's Tortoises (*Gopherus berlandieri*) in southern Texas, females reached maturity at 5 years of age, which is early for the genus. Females of other *Gopherus* species mature at 9–21 years (Gopher Tortoise) and 13–16 years (Mojave Desert Tortoise). These differences match the broader correlation between age at maturity and adult survival rates. Like other vertebrates, turtle life spans are roughly proportional to maturation ages. Longer-lived tortoises (Gopher and Mojave Desert) mature later than shorter-lived tortoises (Berlandier's).

In tortoises, females typically mature later than males. Female Steppe Tortoises grew at the same rate as males but reached maturity later, therefore having a larger body size. The same has been observed for Spur-thighed Tortoises in Algeria and in Spain. Males reached sexual maturity earlier than females, with a variability of up to 3 years. In Yellow-headed Tortoises, males matured at a minimum body size of 175 mm, whereas the minimum for females was 240 mm, with corresponding differences in age at sexual maturity (males 5–6 years versus females about 8 years). So, a tortoise population may frequently contain younger males courting older females.

Environmental factors also influence the age at sexual maturity. For example, populations of Aldabra Giant Tortoises at higher densities, where there is less food for each individual, attain sexual maturity later in both males and females. Resource limitations slow their growth rates and prolong the time to sexual maturity. Growth rates of Mojave Desert Tortoises over more than 40 years were correlated with the amount of preferred vegetation available (softer, ephemeral plants), which varied with winter rainfall (Chap. 6).

Once mature, turtles and especially tortoise appear to remain reproductive for decades. Tortoises have small clutches and spread out their reproductive effort over their long lives. As mature tortoise females grow, they produce increasingly more eggs (because of the constraints of body size on clutch size), at least until growth rate slows to a near stop. With variable and often high rates of hatchling mortality (above), tortoises rely on a long reproductive life span to yield the next few generations of offspring. Their rates of population increase are among the lowest observed in any vertebrate animals.

Slow Aging

Tortoises live a long time, with little physical aging compared to mammals. A study of captive tortoises showed that 80% of species aged more slowly than humans. The larger the body size, the longer-lived the tortoise. Importantly, there is no evidence of reproductive senescence.

How do tortoises defy the inevitable aging (senescence) that happens in other animals? Aging in animals is considered to result broadly from the accumulation of cellular damage over time. There's a suite of molecular hallmarks of aging that add up to damaged cell functions, including mitochondrial dysfunction, toxic proteins, and reactive oxygen products, through accumulation of mutations. The ability to repair damaged deoxyribonucleic acid (DNA) is a demonstrated characteristic of longer-lived animals. As tortoises continue to grow larger throughout their lives, they may be less vulnerable to predators and other hazards, thus allowing them to invest more in cell maintenance to keep their cells functioning well.

Longevity Genes

A study of genomes from Aldabra Giant Tortoises, as well as from Pinta Island's Lonesome George (*C. n. abingdonii*, estimated at more than 100 years old when he died in 2012), revealed activity in a suite of genes related to longevity. These include genes having to do with metabolism regulation, immune response, intercellular communication, and maintaining the right composition of proteins. For example, variants were found in giant tortoises for genes related to tolerating low-oxygen conditions, resisting cancer, and coping with other stressors. The giant tortoise genome even includes the expression of genes that help repair DNA, such as a gene variant also found in the longest-lived rodent, the naked mole-rat. To date, the only fully sequenced giant tortoise genomes are from "Lonesome George" (Galápagos Giant Tortoise) and from "Hermania" (Aldabra Giant Tortoise). Further sequencing of giant tortoise genomes may provide clues to their innate longevity.

Longevity relates to resilience (Chap. 3), with more resilient phenotypes (genetically determined characteristics) likely to get passed on through subsequent generations. For example, genes for withstanding temperature extremes could confer longevity to organisms. In tortoises, a suite of genes has been identified that collectively confer hardiness. Studies of the Mojave Desert Tortoise have found genes associated with regulation of urine volume, response to ultraviolet light, and circadian rhythms. The study of these genes may explain unusual tortoise characteristics, such as their long life spans and ability to survive harsh conditions.

Mammals stop growing and reproducing at specific ages and/or stages; therefore older individuals cannot pass age-deterrent genes to the next generation. Hence there is no natural selection against mutations that reduce human health after reproductive years. In contrast, tortoises lack a definite end to growth and reproduction and are thus subject to stronger selection against mutations that would cause age-related declines. They are a model for a species that grows slowly, reproduces late, and ages slowly. Their physical protection of the shell is also part of the evolutionary package of turtles—and especially tortoises—aging slowly. Vertebrates with protective armor generally age more slowly relative to their body size.

Mortality

Given their natural longevity, factors other than cellular aging are likely responsible for most tortoise mortality. It's difficult to determine the causes of death in wild tortoises, because often the only remains are parts of the shell and other bones. Studies to date indicate that males and females within a tortoise species have similar survival rates; that is, they are equally vulnerable to the many sources of mortality. The positive relationship between body size and survival holds for adult tortoises within a species. In Hermann's Tortoises, for example, larger tortoises survive longer on average.

Causes of tortoise mortality include direct predation and harvest, starvation owing to changing environmental conditions such as loss of habitat, and disease. A major cause is habitat degradation stemming from habitat destruction by humans as well as introduction of invasive species that change availability of forage for tortoises or directly prey on them. You'd think that the thick shell of an adult tortoise would make it invulnerable to predation, but that is not the case, since parts of its limbs remain exposed even when withdrawn (Chap. 2).

Tortoise Predators

When pirates and whalers began visiting the Galápagos Islands during the 17th and 18th centuries, their ships brought rats, dogs, and pigs, which became feral on the islands. The Black Rats (*Rattus rattus*) on Isla Pinzón, for example, quickly multiplied thanks to the plentiful supply of tortoise eggs and hatchlings. Soon there was no recruitment of new tortoises to the population, which was the case for more than a century until the rats were exterminated. Still, other introduced predators continue to have a huge effect on populations of many tortoise species through predation on juveniles and adults. Wild dogs prey on Mojave Desert Tortoises despite their substantial shells.

Although introduced predators tend to have outsized effects on tortoise species that did not evolve among them, native predators also target adult tortoises. Larger mammals like Mountain Lions (*Felis concolor*), Bobcats (*Lynx rufus*), and Coyotes (*Canis latrans*) prey on adult Mojave Desert Tortoises. Jaguars (*Panthera onca*) have preyed on Sonoran Desert Tortoises (*Gopherus morafkai*), and Mountain Lions occasionally kill Mojave Desert Tortoises when their paths cross (which is not often, given different habitat preferences). There is also evidence of bear predation on tortoises, such as Malaysian Sun Bears (*Helarctos malayanus*) consuming Asian Giant Tortoises in Borneo. These examples are all large, mammal predators, but any predator with appendages that can scoop exposed parts of tortoises out of the shell openings—such as American Badgers (*Taxidea taxis*) in the Mojave Desert—may go after adult tortoises.

Some birds have developed a novel way to access the flesh of adult tortoises. The Kelp Gull (*Larus dominicanus*) visits Dassen Island off the coast of South Africa, where it leaves evidence of its predation on Angulate Tortoises in the form of broken and emptied tortoise shells scattered on the granite outcrops. The forensic evidence—clean breaks of the shells, still-attached limbs, and blood stains—points to gulls dropping tortoises on the rocks to break them open and consume their innards. Similarly, South Africa's elusive Karoo Dwarf Tortoises are found smashed on rocks and hollowed out, presumably the leftovers from bird predation. During their reproductive season, Golden Eagles (*Aquila chrysaetos*) in Greece use their talons to grab Hermann's and Spur-thighed Tortoises (in the north) and Marginated Tortoises (in the south). They drop them onto the rocks, sometimes repeatedly, to break them open.

And predation is not necessarily a rare event for adult tortoises, especially the smaller-bodied species. In a population of Speke's Hinge-back Tortoises (*Kinixys spekii*) in Zimbabwe, more than 80% of dead tortoises showed evidence of predation as punctures or other damage to the shell. Southern Ground Hornbills (*Bucorvus leadbeateri*) may hole-punch their shells to pluck out bits of meat (Plate 18).

Disease

Disease is also a factor in mortality of tortoises that can have substantial effects on populations. Tortoises carry a range of pathogens including viruses, bacteria, and fungi, which may be introduced via ticks or other parasites. Speckled Dwarf Tortoises in the Karoo biome of South Africa were found to host 34 bacterial species, a substantial load of fungi and yeasts, cysts from protozoan parasites (*Coccidia*), and nematode worm eggs. Galápagos Giant Tortoises carry both adenovirus and herpesvirus. Whereas many of these organisms and viruses may be commensal (not harming the tortoises), stressors like drought or other climatic factors can change their microbiota and sicken animals. During dry years, the numbers of ticks infesting Speckled Dwarf Tortoises (ticks from the genus *Ornithodoros*) increase.

For the North American tortoises, particularly the western desert species, upper respiratory tract disease (URTD) has caused extensive mortality. Its inflammation of the tissue of the nasal and tracheal area retards breathing and activity. Some tortoises survive the infection, but many become asymptomatic carriers and infect new individuals. The cause has been traced to the bacteria *Mycoplasma agassizii* and its less virulent cousin *M. testudineum*. Ongoing research into the etiology of URTD suggests that co-infection with another bacteria—*Pasteurella testudines*—may exacerbate illness. *Mycoplasma* and *Pasteurella* live side by side without affecting each other, but they cause inflammation in Mojave Desert Tortoises, especially larger (presumably older) ones.

Life History

It's fair to say, then, that most tortoises have life history traits of what's called a *K*-selected species, one in which individuals take a long time to mature, produce relatively few young at a time (in contrast to a marine turtle that lays more than 100 eggs), and live a long time. The tortoise life history strategy has a suite of adaptations for producing generations of young over many decades. Compensating for the high risk of mortality of juvenile tortoises is the extended reproductive capabilities of females.

To sustain a population, a male and female pair must yield just one successful pair of offspring that survive to adulthood and reproduce in kind. Yet most tortoise species are threatened with declining populations, and some are on the brink of extinction. Humans—through intensive harvest and habitat alteration—have managed to back this resilient group of animals into a corner where even two successful offspring per pair are unlikely for many species (Chap. 9).

CHAPTER 6

Deserts to Rainforests

Tortoise Ecology

Tortoises are subtropical to tropical terrestrial animals (see Fig. 3.1). In the Northern Hemisphere, only two species, Spur-thighed and Steppe Tortoises (*Testudo graeca, T. horsfieldii*), have populations living in the western Asian steppes above the winter 0°C isotherm. In the Southern Hemisphere, only a few populations of Chaco Tortoises (Chelonoidis chilensis) live near the winter 10°C isotherm. Most species are lowland to mid-montane (up to 300-m elevations). The Sulawesi Tortoise (*Indotestudo forestenii*) on Indonesia's Sulawesi Island, however, has been found at elevations of 1,000 m, and the tiny Greater Dwarf Tortoise (*Homopus femoralis*) at elevations of 1,900 m in South Africa.

All tortoises are terrestrial. Their family (Testudinidae) contains no aquatic species. Tortoises are commonly associated with semiarid habitats like scrublands or arid grasslands in the Southern Hemisphere. Their suite of adaptations to survive drought and food scarcity (Chap. 3) may allow tortoises to tolerate conditions that many other herbivores cannot, which may reduce competition for food. Still, some tortoises occupy high-humidity habitats like tropical rainforest or wet evergreen forest, such as the Yellow-footed Tortoise (*Chelonoidis denticulatus*; Plate 25) and the Travancore Tortoise (*Indotestudo travancorica*; Plate 21). Asian Giant Tortoises (*Manouria emys*) seek moisture within their already humid habitats, sheltering under rotting logs or other debris and regularly dipping into rainwater pools and small streams.

Keystone Species

Tortoise lifestyles shape their habitats in ways that support other species of animal and plant. As such, they are keystone species with disproportionate influence on the nature of their ecosystems. In some arid areas with less than 50% humidity, tortoises excavate burrows that provide refuge for other animals from the dangers of desiccation and extreme temperatures. Galápagos Giant Tortoises seek wet conditions where they wallow, scouring out muddy depressions in freshwater wetlands that support water-loving plants. Even where they don't burrow or wallow, such as in tropical forests with 80%–90% humidity, tortoises still shape their ecosystems as major consumers of vegetation and dispersers of seeds. In any habitat, tortoises process large quantities of plant material through their digestive systems, which moves nutrients through ecosystems.

Tortoise Diets

Plant Lovers

Tortoises are largely herbivorous, with diets consisting of a variety of plants like a field greens salad. Each species' diet changes with seasonal availability and features diverse plant parts from roots, shoots, leaves, and flowers to fruits and seeds. Juvenile Gopher Tortoises (*Gopherus polyphemus*) in Florida have been documented eating 26 different species of plants; Mojave Desert Tortoises (*Gopherus agassizii*) in Arizona and Utah eating 24 plant species; and adult Leopard Tortoises (*Stigmochelys pardalis*) in Tanzania eating up to 47 species. At least for desert-dwelling tortoises, their diet is dominated by annual plants that don't tend to have a high content of alkaloids, tannins, or other defensive compounds that make them unpalatable or toxic. If given a choice, all tortoises will choose newly sprouted annual plants as well as fresh growth of perennial grasses, shrubs, and trees.

Farm Fresh

Where tortoises live in association with humans, they may also consume cultivated crops. Souss Valley Tortoises (*T. g. soussensis*), which have been displaced from much of their natural habitat in Morocco, feed on the edges of cultivated fields, eating mostly green beans, tomatoes, lettuce, zucchini, cauliflower, and other crops. Yellow-headed Tortoises (*Indotestudo elongata*) in northeast Thailand graze on home garden crops such as jackfruit, mango, and sugar apple (sweetsop). Part of their range is named "Tortoise Village," with tortoises not only tolerated but reportedly fed papaya, star fruit, rose apple, cucumber, melon, pineapple, and other crops.

Picky Eaters

Not all plants are alike from a tortoise perspective. Every species selects certain plant groups, based on what's growing in their habitat (Fig. 6.1). Their diets likely reflect nutritional content, digestibility, and concentrations of toxins. Depending on the seasonal availability of plants, it may be worth foraging for longer to find

Figure 6.1
Tortoises that live in arid landscapes—such as this South African Tent Tortoise (*Psammobates tentorius*)—rely on succulent plants like this fig marigold for nourishment and water.
Photograph by Alexey Yakolev, Wikimedia-CC.By.jpg

higher-value options. For example, Hermann's Tortoises (*Testudo hermanni*) in Montenegro and Croatia favor legumes (family Fabaceae) such as beans, clover, and lupins when they are available; legumes are nutritious and easy to digest. Despite the abundance of harder-to-digest grasses, Hermann's Tortoises and their cousins the Steppe Tortoises rarely eat them. Spur-thighed Tortoises in Algeria also favor legumes, along with crucifers (family Brassicaceae). Of 40 species of edible plants growing in the Mergueb Nature Reserve (Algeria), Spur-thighed Tortoises ate only 11 species during an April observational study, likely reflecting a preference for certain nutrients (e.g., carotenoids, vitamins, minerals, and the high nitrogen content of legumes).

In some areas with fewer options, grasses may be a valuable part of tortoise diets. For example, the grasses of the Karoo habitat of South Africa are a dietary staple for the Leopard Tortoise. On the Aldabra Atoll, some 20 species of grasses and herbs make up the so-called tortoise turf, favored by the Aldabra Giant Tortoise (*Aldabrachelys gigantea*; Plate 9). In general, tortoises choose more specialized diets, with a focus on nutritious legumes in moister habitats. In dryer, less vegetated habitats, tortoises become generalists and include less nutritious vegetation simply because there are fewer options.

To Each Its Own

Even individual tortoises within a species do not necessarily pursue identical diets. A study of Mojave Desert Tortoises discovered a surprising amount of individual variation in food preferences. All the tortoises favored a diet of plants rich in protein but low in fiber, presumably because they need the maximum digestible energy. But individuals of the same species of tortoise preferred different suites of plants. The broader the food options, the more individual preferences may diverge, which spreads tortoises out into the available feeding "space."

There are frequently age-class dietary differences as well. In Mojave Desert Tortoises, juveniles have more specialized diets than adults, reflecting their more limited movements, lower height reach, reduced gut capacity, and differing metabolisms. Juvenile Aldabra tortoises feed almost entirely on the succulent herbs growing amid the grasses of the "tortoise turf." As tortoises mature physically, their diets may shift or broaden.

Meat on the Side

No tortoise species is strictly vegetarian. Although they are mostly herbivorous, tortoises periodically include living and dead animal protein (Fig. 6.2) and other

Figure 6.2
A Gopher Tortoise (*Gopherus polyphemus*) feeds on a discarded fish carcass to get a bit of protein, calcium, and likely salt.
Photograph by Ken Dodd

nutrients in their diets. Given the opportunity, they will eat arthropods and snails, and scavenge rotting vertebrate carcasses. Indeed, their consumption of meat may be largely limited by availability. A study of Travancore Tortoises revealed diets consisting of only 45% plants, with the remainder consisting of carrion and living animals such as crabs and millipedes. Radiated Tortoises (*Astrochelys radiata*) in southern Madagascar have been observed feeding on mammal and fish carcasses, feces, and bones; in Aldabra, tortoises commonly feed on the carcasses of other tortoises. Southeastern Hinge-back Tortoises (*Kinixys zombensis*) in Madagascar forage for snails and discarded meat such as chicken gizzards in gardens and rubbish piles. Rarely, tortoises catch and eat vertebrates; for example, an introduced Aldabra Giant Tortoise caught and ate a fledging Lesser Noddy (*Anous tenuirostris*) on Fregate Island, Seychelles.

Hinge-back Tortoises (*Kinixys* spp.) from southern Africa are the most omnivorous tortoises, mixing a diet of plants, mushrooms, and invertebrates. Mushrooms supply the most digestible energy, but invertebrates provide nutrients like minerals and amino acids to build proteins. Many *Kinixys* are reported to eat millipedes, apparently biting them right behind the head without any ill effect from the millipedes' defensive secretions. In a Zimbabwe population of Speke's Hinge-back Tortoise (*Kinixys spekii*), invertebrate prey was nearly a fifth of tortoise stomach contents, and millipedes by volume were 65% of those invertebrates. Although the mechanisms are unknown, Hinge-backs apparently have a high tolerance to the toxins, which include a variety of noxious compounds such as benzoquinones, hydrogen cyanide, and hydrochloric acid. Millipedes are relatively low in calories and take a long time to digest but offer Hinge-backs almost 500 times as much calcium as mushrooms.

Snacking on Scats

A common and easily obtainable source of additional nutrients for tortoises is feces ("scat"). The Angulate Tortoise (*Chersina angulata*), for example, eats rabbit feces during summer and fall when food resources become scarce. Hermann's Tortoise in Serbia is also known to ingest mammal feces, a departure from herbivory that may satisfy its needs for protein and calcium. Ploughshare Tortoises (*Astrochelys yniphora*) have been observed feeding on dried African bush pig feces. And Leopard Tortoises reportedly gnaw on animal bones and eat hyena feces. In a two-for-one, Spider Tortoises (*Pyxis arachnoides*) eat cow dung infested with insect larvae, so they get the nutritional benefits of both the feces and the larvae.

There are also instances of turtles eating the feces of their own species, which may confer not only nutrition but also a gut flora (intestinal microbes) for effectively digesting plant fibers made of cellulose (Chap. 3). We have found three records of same-species feces consumption in tortoises: Red-footed Tortoises (*Chelonoidis carbonarius*) consume feces of other Red-foots; Aldabra Giants consume feces of fellow Aldabra Giants; and Mojave Desert Tortoise juveniles eat feces from adults of their species. Although feces eating may be widespread, tortoises still leave a high volume of undigested plant matter in their feces.

Mushroom Lovers

For tortoises that live in sufficiently humid climates, mushrooms are a staple food during seasonal rainy periods. Twenty species of tortoise across five continents (Africa, Asia, Europe, North America, and South America) have been documented eating fungi. The Impressed Tortoise (*Manouria impressa*) of Southeast Asia specializes in mushrooms (Fig. 6.3) and other types of fungi in its mountainous forest habitat. During its active foraging season in the spring and summer, it's known to consume at least eight mushroom species and little else other than occasional bamboo shoots and insects. Its major season of activity corresponds to the warmer, rainy season when mushrooms appear. Tortoises may obtain soil minerals and concentrated energy through fungi, as well as hydration, since fungi consist of up to 90% water.

Getting Their Minerals

Tortoises need specific nutrients to produce and sustain bones, shells, and eggs: phosphorus, magnesium, sodium, and particularly calcium for females that are shell-

Figure 6.3
Sketch of a juvenile Impressed Tortoise (*Manouria impressa*) eating a mushroom, a preferred food for it and many other species. Sketch derived from purchased iStock photograph

ing up their eggs. Western *Gopherus* actively eat caliche—a sedimentary rock made of calcium carbonate. Female Mojave Desert Tortoises have been observed scraping off topsoil with their sharp beaks to access the mineral-rich "caliche subsoil" underneath (Plate 6). Their ingestion of the calcium carbonate–rich caliche is closely associated in time with females' egg production and shelling of the ova (Chap. 4). At least in some species, males "eat dirt" (geophagy) as well, perhaps to supply mineral needs. Hermann's Tortoises ingest both dry soil and mud.

Adding Some Salt

Salt is a limiting nutrient for animals that are mostly herbivorous. African Spurred Tortoises (*Centrochelys sulcata*) have extended home ranges that include a "salt lick" to which they travel periodically to supplement their naturally low-salt herbivorous diet. Similarly, Leopard Tortoises in Zimbabwe were observed frequenting an area of high-sodium soil, which might account for their unusually large home ranges. Their long-distance beelines of dozens of meters to the salty soil may make up for the low sodium content of the plants they eat.

Tortoises must also adjust their diets to avoid excess intake of certain nutrients. In arid environments, tortoises often avoid plants that contain potassium. Potassium ions are retained in the urinary bladder and, if concentrated, can interrupt cell and nervous system functions. Tortoises remove excess phosphorus by combining it with nitrogen in their bladders to produce potassium urate crystals, which are not toxic (Chap. 3).

Seasonal Variations in Diet

Tortoise diets vary seasonally in tandem with the availability of different resources. Angulate Tortoises on Dassen Island, South Africa, rely on herbs and seedlings during the wet season, and on dried plant materials during the dry season. Leopard Tortoises adjust their diets seasonally with shifts in the relative abundances of grasses, succulents, herbs, and fruits. Berlandier's Tortoises (*Gopherus berlandieri*) live in semiarid habitat with abundant patches of Prickly Pear (*Opuntia* sp.; Fig. 6.4) and eat the spiny cactus pads ("nopales") year-round, but in late summer they feed exclusively on the ripened, bright-red fruits ("tunas"), which are rich in water and nutrients.

Seasonal changes in diets of tortoises do not always vary with plant abundance. Experiments with Mojave Desert Tortoises showed that they frequently include rare species in their diet of forage plants. Out of the hundred or so available plants, this species' diet consists mostly of their top five preferred plant species throughout their annual activity period. The preferred are largely the plants that remain green and

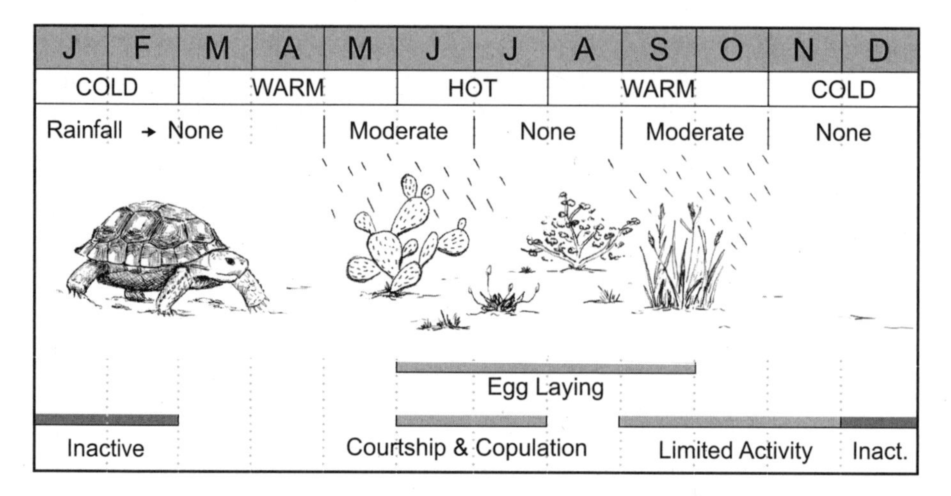

Figure 6.4
A Berlandier's Tortoise's (*Gopherus berlandieri*) year in profile displaying the major periods of activity and inactivity associated with normal weather conditions. Concept from Esque et al. (2014); data from Auffenberg and Weaver (1969) and Rose and Judd (2014)

edible for the longest part of their activity season, obviating the need for tortoises to shift foods.

The energy needs of mature female tortoises tend to fluctuate seasonally and peak as their bodies physiologically prepare for egg laying (Chap. 5). Their diets typically expand during the high-energy-use periods. Although Hermann's Tortoises rarely feed on carrion, a female was observed scavenging a long-dead Common Toad (*Bufo bufo*) during her nesting season. Aside from offering minimal nutrients, it might expose her to poisonous skin toxins, suggesting she was desperate for food.

Shifting Grocers

In some areas, tortoises deal with seasonally changing food supplies by migrating. Female Sonoran Desert Tortoises (*Gopherus morafkai*) move seasonally to higher elevations where they encounter greater diversity of plants on the northern slopes after summer rains. The majority of Western Santa Cruz Giant Tortoises (*Chelonoidis niger porteri*) migrate between the lowlands and highland each year. On the Galápagos Island of Santa Cruz, the highlands stay humid during the June–December dry season because of daily, moisture-heavy, thick fog. Tortoises migrate to the highlands and feed on grasses and small herbs, then return to the warmer lowlands in January as the lowlands become green with vegetation from seasonal rains. They leave well-worn trails (Plate 23) from their years of migrating along the same paths.

There's a significant energy cost of migrating to the highlands, with individuals traveling 10 km or more from sea level to over 400 m in elevation, which can take up to three weeks. Studies of body condition show the benefits of migrating for the dry season; tortoises in the highlands have higher protein and red blood cell counts than those that don't migrate. Still, food quality does not entirely explain the seasonal migration of Western Santa Cruz Giant Tortoises. Curiously, their migration timing does not always track closely with seasonal weather changes. For example, an unseasonably dry year did not cause an earlier migration despite the earlier drying of the lowland vegetation, indicating that other factors also prompt movements to the moister uplands.

Summer and Winter Homes

Other seasonal changes in habitat, such as microclimate and availability of standing water, play a role in the habits of Galápagos (Fig. 6.5) and other tortoise species. Hermann's Tortoises in Italy select distinct habitats during their active warm season versus their inactive winter season. In summer, they favor the low vegetative cover of

Figure 6.5
Galápagos Tortoise (*Chelonoidis n. porteri*) on Santa Cruz Island soaking in a pool of water. Immersion in water likely helps with thermoregulation and hydration, although that has not been experimentally tested.

barren lands, open shrublands, and open forests, where they take refuge in brambles. For winter hibernation, they select dense forests and shrublands with greater leaf litter for insulation. During the spring breeding season, they favor habitats with new herbaceous growth that is easily digested.

Juvenile tortoises may show seasonal patterns distinct from those of adults. In Mojave and Sonoran Desert Tortoises, for example, juveniles are active on warm winter days when adults are not. The most active juveniles occupy shorter burrows and are thus more directly exposed to daily fluctuations in temperature (see the Burrows section). Juveniles also get more active than adults during winter warm-temperature spikes in other tortoise species, such as Gopher Tortoises and Egyptian Tortoises (*Testudo kleinmanni*), where juveniles come out to bask at the mounded entrances of burrows on warm days. The small body size of juvenile tortoises allows them to heat up quickly and remain active in winter.

Home Ranges

All tortoises conduct their activities within specific parts of their habitat. The total area encompassing their activities defines their home range, including locales where a tortoise shelters and regularly forages for food and water. Tortoises have lower energy needs than mammals because tortoises do not maintain consistently high metabolisms (Chap. 3), yet their home ranges may be of comparable size. Because resource needs determine home ranges, tortoise home ranges naturally vary by species, sex, age, habitat quality, season, and weather. Consequently, they can be highly variable (Table 6.1).

The features of a habitat—both biotic and physical—affect home range size because a tortoise's home range must satisfy its various needs. For African Spurred Tortoises in the dry West African Sahel, the home ranges correlate with the presence of trees and shrubs and the absence of bare areas ("zipele"). Human settlements may also shape tortoise home ranges; African Spurred Tortoise ranges rarely include areas near villages, although this pattern may stem from direct harvest of tortoises by villagers.

Home range size can also vary as a function of body size of tortoise species. A study that followed tortoise activity for a week in Zimbabwe calculated an average home range size of 26 ha for Leopard Tortoises compared to an average of only 1.9 ha for the smaller Speke's Hinge-back Tortoises. The still smaller Hermann's Tortoise has an estimated weekly home range of about 0.5 ha. A smaller body size implies lower food intake, which may be satisfied by a geographically smaller foraging area, although distinct nutrient needs of species play a role as well.

Table 6.1.
Home Range Areas for a Variety of Tortoise Species Living In Different Habitats

Standard Name of Tortoise Species	Home Range(ha)	Carapace Length (cm)	Geographic Location
Spur-thighed	0.4–8.9	~5–23	west-central Morocco
Hermann's	~8	10–15	southeastern France
Geometric	0.35–34.0	10–14	Western Cape, South Africa
Impressed	2.7–17.7	17–30	northeast Thailand
Aldabra Giant	4.9	adults	Malabar Island, Aldabra Atoll
Western Santa Cruz Giant	2.9	50–150	Isla Santa Cruz, Ecuador
Mojave Desert females	12–22	17–38	Nevada, USA
Mojave Desert males	15–72	22–37	Nevada, USA
Sonoran Desert females	1.8–19.6	adults	Arizona, USA
Sonoran Desert males	10.6 ± 4.5	adults	Arizona, USA

Note: The areas are calculated by different computational methods; nonetheless, the areas reported herein are roughly equivalent and represent the space used on daily forays for food and shelter from a fixed center of activity.

Measuring Home Ranges

Distinct methods for measuring home range offer the potential to corroborate home range estimates. The minimum convex polygon (MCP) method or kernel density estimator (KDE) are commonly used alone or in combination. The MCP may better capture the total area used by a tortoise (because it connects the dots for the farthest points of its range), although the KDE highlights areas of intensive use. Both measures are subject to variation depending on how the home range boundary is adjusted across sampling points.

Studies of activity patterns over shorter time frames, such as weekly or monthly, yield smaller and less variable home range sizes, with occasional outliers (Table 6.1). Leopard Tortoises in Swaziland, for example, typically move 89 m per day but have a maximum daily movement of more than 300 m that would not be detected in a shorter study. Tortoise home range estimations, like estimations of other animals' ranges, should err on the side of ample sample points over long periods and repeat studies with increasingly accurate measurement technologies applied as they become available.

In sum, the variable home ranges and movement patterns reported in tortoise studies reflect differences not only in the biology and sex of tortoises and weather conditions, but also in the definition of terms, time frame of study, and manner of data collection.

Sex Effects

Males and females of many tortoise species have different home range sizes, affected by females seeking foods to support egg production and males seeking females to mate. Usually, the home ranges of groups of tortoises abut or overlap slightly, and these clusters of individual home ranges establish a community in which social interactions occur, mostly around reproductive activity. Females in some species of tortoises consistently have larger home ranges than males, a difference that may reflect distinct dietary needs. In Geometric Tortoises (*Psammobates geometricus*) in South Africa, the larger-bodied females have bigger home ranges than males, as do Steppe Tortoises in Uzbekistan, where female home ranges average more than twice the area of male ranges. Examining how they use the space reveals that females walk wide loops around their larger home ranges, whereas males intensively patrol smaller areas. Females may range more widely to find dietary resources or to evaluate potential mates, and male ranges may be restricted by the size of territory they're able to defend.

Larger female home ranges are not the rule, however. In some tortoise species, such as Red-footed Tortoises in the Bolivian Chaco, the males range more widely, using about three times as much space as females. A similar pattern has been reported for Mojave Desert Tortoises, perhaps driven by males wandering in search of females during the mating season. Outside of mating season, home ranges of tortoises tend to shrink. In Steppe Tortoises and other species, males stop patrolling territories and devote more time to foraging.

Juvenile Movements

Juvenile tortoises typically have smaller home ranges than adults. Juvenile Mojave Desert Tortoises remain around a single set of burrows, in contrast to adults—especially males—who move between distant sets of burrows that serve as centers of activity. Such differences may stem from their distinct nutrient requirements or other factors like the mating-driven forays of adult tortoises. Also, juvenile tortoises typically have unique structural resource needs, such as the cover of creosote bushes and the small mammal burrows used by juvenile Mojave Desert Tortoises (see the Burrows section).

You would expect juveniles—because of their vulnerability to heat stress and predators—to spend substantially more time sheltered than adults, leading to fewer movements and smaller home ranges. Notwithstanding, recent movement data using Global Positioning System (GPS) technology showed three- to seven-year-old Gopher Tortoises leaving their burrows more often, ranging farther, and spending longer periods away from burrows than previously observed. Such data reinforce the difficulties of monitoring small, cryptic animals and the variability introduced by available technology.

Habitat Quality

Home range sizes of tortoises correlate with habitat quality. As the density of resources increases, the amount of space a tortoise needs to meet its resource needs decreases. It follows that within a species, home range size may vary across populations that live in different habitats. In northern Italy, for example, the home ranges of Hermann's Tortoises in a wooded habitat with scarce food supplies were many times larger than the typical home ranges for the species. Individual tortoises were more spread out, as you might expect in resource-scarce situations. Similarly, the home ranges of populations of Leopard Tortoises in South Africa scale with habitat quality; in the driest Nama-Karoo habitat, tortoises move up to a record 8 km per day (although typically several kilometers), compared to tortoises in valley thicket (up to 100 m/day) and in the lushest Swaziland habitat (just 50 m/day). Gopher Tortoises in resource-poor southern Florida scrub habitat have larger home ranges than Gopher Tortoises in higher-quality sandhill habitat.

Seasonal Effects

As habitat quality changes seasonally or stochastically, tortoises adjust their home ranges to ensure access to all needed resources, including food, water, shelter, mates, and warmth. Speckled Dwarf Tortoises (*Chersobius signatus*) have relatively small home ranges, their needs met by the abundant, seasonal plant growth from spring rains in their southern Africa Karoo habitat, but they nearly triple their home ranges during a drought year. Similarly, Geometric Tortoises have larger home ranges during the dry season when plants are less verdant. During the wet season, they forage in a smaller area and may be restricted in their movements by temporary pools of standing water that they won't cross. Other geographic features can constrain movement and therefore limit home range size; Sonoran Desert Tortoises living in rugged terrains such as mountain passes have smaller home ranges than those inhabiting alluvial basins (bajadas) and valleys.

Individual Variation

Even within a species, not all tortoises make the same home range choices. Within Aldabra Giant Tortoise populations, there are sedentary individuals that move little even if food is scarce, whereas most of their "compatriots" make seasonal shifts, moving from grasslands to the adjacent open scrub forest. There is a third group of individuals that favor woodlands, moving at the beginning of the rainy season to coastal grasslands and then returning to the forest prior to the dry season. Among other features, tortoises of this subpopulation have proportionately narrower and longer shells than the sedentary grassland tortoises, suggesting some localized genetic change.

Galápagos Tortoises also exhibit dual strategies that vary by individual tortoise. Typically, migratory individuals are capitalizing on resources that change seasonally in predictable ways. For example, all tortoises living on Volcán Alcedo (Isla Isabela)—where conditions vary the most over time—are "migrators." Every year, they make the round-trip journey from dry lowlands to humid highlands and back. In contrast, on Isla Española, where conditions vary less, more than 80% of the tortoises remain in the same area year around as "residents."

Leopard Tortoises in South Africa have distinct individual strategies that are not seasonal. Tortoises that live near standing water make regular journeys to drink water. Individuals living farther away rely on succulent plants for water and thus move less frequently and over shorter distances.

Core Areas

Tortoises with broad home ranges usually concentrate their activity in a subset of spots—core areas—that meet their daily needs for resources. Forest-dwelling species of Hinge-back Tortoises in Cameroon—Forest Hinge-back Tortoise (*K. erosa*) and Home's Hinge-back Tortoise (*K. homeana*)—spend most of their time in core areas that comprise less than 25% of their home ranges. Even though the home ranges of individual Hinge-back Tortoises overlap, their core areas do not overlap, indicating habitat partitioning. In general, tortoises don't defend territories with fixed boundaries like birds or some mammals, although they divide up preferred habitats in ways that reduce overlap and maximize their individual access to resources.

Singular assets such as burrows may be the linchpins of a core area. Mojave Desert Tortoises concentrate their activities around one or more burrows, with periodic forays to other sets of burrows during the warmer, active season. Some species of tortoises show site fidelity by consistently using the same burrows from year to year. Sonoran Desert Tortoises, for example, emerge from their burrows every spring, use other areas during summer rains, and then return to the same burrows every fall. Thus an individual tortoise's spring and fall home ranges from year to year are highly consistent, with a redundancy of about 80%. They navigate back to known areas using a combination of sight and smell (Chap. 3).

Ecological Engineering

Pruning Machines

Like elephants, tortoises are ecosystem engineers. Simply through their consumption of vegetation, tortoises can substantially alter the plant diversity and coverage of their landscapes, for better or for worse. On the extreme end are giant tortoises, the largest herbivores in the Galápagos and Seychelles. They are estimated to eat as much as 11% of plant primary production (new growth) in their habitats. Their large and heavy bodies also break up vegetation, create mud wallows, and abrade rocks as they lumber through the terrain. Galápagos Giant Tortoises affect wetland vegetation as well, as they wallow in ponds or lakes in the highlands of some islands, grazing on aquatic plants.

Absent pressures from farming and other land use changes, tortoise populations can become surprisingly dense. Leopard Tortoises were reported to reach a density of 85 individuals per square kilometer in the Eastern Cape of South Africa (about the density of people in Spain). Just as the density of humans can exceed the available resources, so can the density of tortoises, especially for scarce resources. For example, the grasslands of Aldabra are closely cropped by the tortoises only in the vicinity of shade trees that offer daily protection from the sun. The density of tortoises seeking shade has damaged and in many cases killed the trees (Plate 10). The tortoises who don't find shade risk death from overheating.

Good for Plant Diversity

Although intensive herbivory eliminates individual plants, tortoise grazing can increase plant species diversity over time, substantiating their roles as keystone species. By reducing the density of their preferred plant species, foraging by tortoises allows other plant species to gain a foothold. In experimental plots that excluded Mojave Desert Tortoises for two years, several rare desert plant species appeared that were absent in control plots where tortoises foraged. In a different habitat, the Longleaf Pine forests of Alabama, the exclusion of Gopher Tortoises from plots similarly reduced plant species diversity while encouraging more species-poor plant cover.

Indeed, some strategies to control invasive plants capitalize on foraging tortoises. A program that introduced Aldabra Giant Tortoises and Radiated Tortoises to the island of Mauritius found that their grazing better controlled non-native vegetation than weeding, especially given that they favored invasive species, thereby significantly reducing their density, height, and seed production. Reintroductions of giant tortoises that wallow in wetlands may also reshape the distribution of freshwater plants on islands. Reductions in Galápagos Giant Tortoises on Isla Santa Cruz by

17th- and 18th-century sailors caused freshwater wetlands to fill in and become denser bogs, leading to extinctions of some species of wetland plants and rarity of others.

Dispersing Seeds

The effectiveness of animals as dispersal agents for plant seeds depends on how many seeds they carry, how far they carry them, and how the seeds fare when deposited. On all counts, tortoises rank as significant seed dispersers. They spread seeds by eating fruits and then defecating seeds elsewhere, or by carrying seeds stuck to their feet, legs, or shell.

Of the 47 known tortoise species today, nearly half (21 species) consume fruits. Frugivorous (fruit-eating) turtles vary in the degree of specialization on fruits; tortoises tend to be generalists, eating a variety of fruits along with other plant materials. Yellow-footed Tortoises in Brazil, for example, have seeds from at least 25 plant species in their guts at any given time, and consume seeds from about 100 plant species annually. Because tortoises are toothless, their ability to process seeds before swallowing them is minimal, and the seeds appeared to suffer little damage from their journey through Yellow-footed Tortoises' guts.

Natural Germinators

There's evidence that after passage through tortoise digestive systems, seeds germinate better, suggesting a potential coevolution between tortoises as dispersers and their host plants. Seeds of the Prickly Pear cactus in Texas had higher germination rates after passing through the guts of Berlandier's Tortoises. Gopher Tortoises in Florida effectively facilitated the spread of Cocoplum (*Chrysobalanus icaco*) by eating the fruits; seeds found in tortoise feces germinated sooner than others. The landscape-level consequence was higher densities of Cocoplum bushes near tortoise movement pathways. Studies of a critically endangered plant—*Syzygium mamillatum*—on the island of Mauritius revealed that although gut passage of seeds through Aldabra Tortoises did not increase germination rates, it resulted in taller, healthier seedlings.

Following the Fruits

The spatial memory of tortoises shapes their role in seed dispersal. Tortoises remember the locations of food supplies. This memory presumably allows them to track the availability of fruits and adjust their movement patterns to favor locations with fruiting plants. Red-footed Tortoises, for example, seasonally travel to places where fallen fruits will be available. Travancore Tortoises congregate under fruiting trees to capitalize on the food resource. Giant tortoises (from both Galápagos and Aldabra) follow the same movement patterns in the landscape year after year (Plate 23), likely relying on spatial memory for key resources such as fruits.

Even though tortoises travel slowly, the distances they travel may be substantial within home ranges that sometimes exceed those of similar-sized mammals. Food has a long transit time of several days to two weeks through a tortoise digestive system (Chap. 2), so seeds are often dispersed a long way from their parent plants. Yellow-footed Tortoises in Brazil carried seeds an average of 277 m (two and a half football fields) during their foraging activities during the rainy season. Red-footed Tortoises in captivity kept seeds in their guts for at least 3 and as many as 28 days, which would give a wild tortoise plenty of time to spread seeds around. Tortoises are considered the main dispersers for many plant species, including Prickly Pear (*Opuntia* spp.), Wild Potato (*Prosopanche americana*), and Wild Tomato (*Solanum esculentum*) in the Galápagos.

Loss of Keystone Tortoises

Extinctions of tortoises can substantially alter plant communities. The extinction of two giant tortoise species—Madagascar Giant Tortoise (*Aldabrachelys abrupta*) and Grandidier's Giant Tortoise (*A. grandidieri*)—may have caused ecological changes, including the distribution of the endemic baobab trees. The baobabs, such as *Adansonia rubrostipa*, have large seeds that would likely have been dispersed in the past by giant lemurs, elephant birds, and tortoises, all of which are extinct. Now, with no known dispersers, Madagascar's large-seeded plants are imperiled, and resource managers have proposed reintroducing giant tortoises to the island (Chap. 10). Because oceanic islands have simpler ecological communities, there may be less redundancy in seed dispersal roles, making tortoises even more vital as a keystone species.

Fire

Some tortoises prefer open habitats that are maintained naturally by fire. For example, Gopher Tortoises inhabit southeastern coastal plains and build their burrows in spots with low canopy cover and high densities of nutritionally rich nitrogen-fixing plants (legumes). Fire naturally reduces canopy cover and promotes legume growth. Regular fires may also control parasites such as ticks, nematode worm eggs, or bacterial cysts. Suitable Gopher Tortoise habitat has become increasingly scarce owing to upland development and fire suppression, evidenced by their decline in longleaf pine stands where fire has been excluded. Gopher tortoise recovery plans therefore include regular prescribed burns to mimic the fires of the past and maintain optimal tortoise habitat.

Tortoises may also play a role in natural fire regulation by eating lots of vegetation, which reduces the accumulation of plant litter. Prehistoric giant tortoises in Madagascar likely helped reduce the intensity and frequency of fires through their grazing. As large herbivores declined after humans arrived, fire frequency increased, as evidenced by charcoal particles in the sedimentary record.

Although a natural fire regime sustains and replenishes tortoise habitats in the long run, the ecological relationship of tortoises to fire is complex. In the immediate term, they are vulnerable to fire because they move slowly and may not escape the flames. With natural fire regimes increasingly disrupted, wildfires are larger and hotter, perhaps presenting tortoises with conditions they did not face in the past. Tortoises may also suffer from the immediate aftermath of fires if habitat burns to the ground, leaving little cover or shade. A study of Spur-thighed Tortoises in northwestern Africa comparing their distribution in recently burnt and unburnt habitats found the density of tortoises reduced by half in the burnt areas. The authors attribute the difference to directly mortality from fires (and slow replacement rates; Chap. 4) as well as degraded postfire habitats with limited shrub cover.

The positive and negative balance of the effects of fire on a tortoise population will depend on the species and the intensity of the fires. Spur-thighed Tortoises prefer habitats with some grass cover, whereas Gopher Tortoises gravitate to open areas. These differences will shape their individual and population-level responses to fire. Immediately after a fire, a burned habitat may be unsuitable in terms of providing shade, but it then becomes richer in plant species as it recovers. When patches of fire are small, tortoises may be able to disperse from the burned area and then recolonize it after fire-adapted plants have sprouted.

Pallets

One of the most dramatic ways that some tortoises engineer their environments is through excavating refuges. All of the smaller and many of the medium-sized tortoise species dig shallow depressions (pallets), usually next to a tree, shrub, or large rock. Pallets offer some protection from predators but largely function to moderate temperatures and limit exposure to sunlight, which reduces the risks of overheating and dehydration. Pallets may be especially important for species that live in areas with scarce vegetation. Studies of the Berlandier's Tortoise show that a tortoise's pallet is cooler than nearby substrates in the sun.

Even giant tortoises dig pallets. The Aldabra Giant Tortoise will sometimes "dig in" once it leaves its sunny grazing area to seek shade beneath a tree. It makes scratching movements with its front feet and forearms while twisting its body from side to side. These movements slowly shift the soil to the side and create a pallet just deep enough (4–10 cm) to expose the cooler subsoil that helps the tortoise keep cool.

Burrows

More than half of the living species of tortoises live in hot landscapes. To take refuge from heat, some dig burrows (tunnels) in areas with sand or sufficiently crumbly soils. Field measurements show that deeper burrows have more stable temperatures, providing a buffer from seasonal temperature extremes and reduces the stress of surface living. Bolson Tortoises (*Gopherus flavomarginatus*) use their burrows primarily based on their inside temperatures; burrows with plant cover tend to be cooler and are more frequently used. Only a few species dig burrows sufficiently deep and long to create a stable microclimate, however. The size and shape of tortoise burrows are species specific and related to the interaction of body size with habitat features. The smallest tortoises are not true burrowers.

All species of *Gopherus* except the smallest (Berlandier's Tortoise) excavate burrows. The Bolson Tortoise and Gopher Tortoise are the strongest burrowers, capable of digging burrows 10 m long and up to 6 m deep for the latter species (Fig. 6.6), although 2- to 6-m-long burrows are the typical length for both species. The burrows of the Mojave Desert Tortoise and Sonoran Desert Tortoise tend to be shorter;

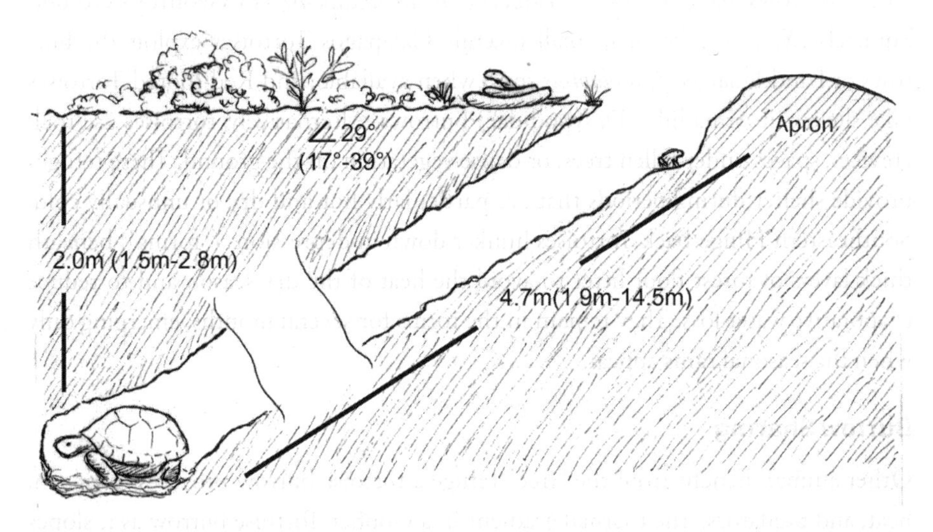

Figure 6.6
Schematic depiction of a Gopher Tortoise burrow showing the range of depths, lengths, and inclinations discovered in research on natural burrows in Florida and southern Georgia, USA. Data mainly from Eisenberg (1983) and Hansen (1963)

both species also colonize natural crevices in rocky landscapes and arroyos. African Spurred Tortoises appear to select sloping landscapes for their burrows, as presumably it's easier to start digging horizontally and then angle downward than to excavate vertically. When complete, their burrows are huge—up to 15 m long—in the Sahel semidesert landscapes where vegetation cover is scarce.

Tortoises construct burrows using their front legs. One leg is extended forward in front of the head, with the stout claws embedded in the soil and then swept sideways and backward, moving the soil with it (Plate 22). This process alternates from left to right, with the inside edges of the forearms pushing the loosened soil away. The tortoise's hind legs push its body forward, which twists it from side to side. Periodically, the tortoise reverses its position and uses the front of its plastron to shove the loosened soil to the surface into a soil heap ("apron") at the mouth of the burrow. The burrow is slightly higher than the tortoise's shell and nearly double its width, so the tortoise can swivel around in its burrow.

Burrow Squatting

Even though tortoises that live in more humid habitats do not excavate burrows, they may still use burrows of other animals or other types of refuges for protection from predators or to thermoregulate. Berlandier's Tortoises commonly inhabit the brush pile nests of Wood Rats (*Neotoma* spp.); Souss Valley Spur-thighed Tortoises co-opt burrows made by Honey Badgers (*Mellivora capensis*) or Geoffroy's Ground Squirrels (*Xerus erythropus*); small juvenile Galápagos Tortoises exploit the burrows of Land Iguanas (*Conolophus* spp.) when available; and Red-footed Tortoises take shelter in Armadillo (Dasypodidae) burrows. Alternatively, tortoises use rock crevices, spaces under fallen trees, or dense vegetation on the ground. These refuges provide shelter during periods that are particularly hot and dry or unusually cold. Southeastern Hinge-back Tortoises hunker down in depressions ("forms") beneath thick layers of forest floor litter to avoid the heat of the dry season and to reduce evaporative water loss. They remain in the forms for several months, emerging only when the seasonal rains appear.

Burrow Sharing

Other animals benefit from the "free" refuge a tortoise burrow provides from sun, heat, and predators. The thermal gradient in a Gopher Tortoise burrow as it slopes gradually deeper into the ground (Fig. 6.6) provides a range of humidity and temperature choices. Some animals, such as the Florida Gopher Frog (*Lithobates capito*), depend on the burrow not only for moisture, but also to feed on the invertebrates that live off tortoise feces in the burrow; thus they prefer active tortoise burrows.

Various snakes use Gopher Tortoise burrows, including the rare Indigo Snakes (*Drymarchon couperi*) that hibernate in the burrows during the winter, often returning to the same burrow across years. Mammals—such as Cotton Mice (*Peromyscus gossypinus*) and Cotton Rats (*Sigmodon hispidus*)—take shelter in Gopher Tortoise burrows, too. Larger animals—Eastern Cottontail Rabbits (*Sylvilagus floridanus*), Common Raccoons (*Procyon lotor*), Burrowing Owls (*Athene cunicularia*), and Virginia Opossums (*Didelphis virginiana*)—capitalize on vacant burrows for nesting, enlarging them as needed. The smaller burrows excavated by hatchling tortoises have correspondingly smaller burrow associates, such as immature toads and snakes.

All told, Gopher Tortoise burrows are used by more than 60 vertebrate species, including snakes, frogs, doves, owls, rabbits, skunks, and coyotes. If you include invertebrates, Gopher Tortoise burrows are used by more than 300 species. Some invertebrates are burrow endemics, living nowhere else. Many birds, although they may not enter burrows, use the burrow aprons for taking dust baths, foraging, and displaying to other birds; these include Savannah Sparrows (*Passerculus sandwichensis*), Northern Mockingbirds (*Mimus polyglottos*), and Eastern Bluebirds (*Sialia sialis*). Because of the value of their burrows to other animals, Gopher Tortoises are considered a keystone species within the ecosystems they inhabit. The density of their burrows is positively correlated with the diversity of vertebrates and in fact is the primary determinant of biodiversity in longleaf pine savannas. As habitat loss continues, populations of other animals that depend on Gopher Tortoise burrows are expected to decrease as well (Chap. 9).

Although their burrow systems are the most studied, Gopher Tortoises are not the only ones to attract other species. Their Bolson Tortoise cousins have burrows more than 1.5 m deep and up to 5 m long, leaving ample space for sharing, such as with Black-tailed Jackrabbits (*Lepus californicus*) and Kangaroo Rats (*Dipodomys* spp.). The burrows may therefore be a magnet for predators. At Bolson Tortoise burrows in the Mapimi Biosphere Reserve of Mexico, coyotes, badgers, skunks, and foxes hunt near tortoise burrows, with at least one observed instance of predation when a coyote snatched a rabbit from a tortoise burrow.

Social Structure

Tortoise pallets, burrows, and other refuges spatially distribute tortoises over a landscape, playing a role in social structuring of their populations. For example, studies of Mojave Desert Tortoises show that they're usually solitary in burrows, with each tortoise using its own set of several burrows. The pattern resembles what's been observed for other *Gopherus* species (Gopher Tortoises and Bolson Tortoises) and, along with incidences of apparent burrow guarding, suggests a burrow territoriality.

Seasonal Socializing

During the active spring months of May and June, an individual tortoise switches between its several burrows. Then, burrow occupancy changes again during the breeding months (July, August, and September). Males roam around more, courting females and living in their burrows for short periods. Each male tends to visit several females over the course of the reproductive season. Females exhibit a strong male bias in their burrow sharing, whereas males share burrows with either sex. By the end of September, tortoises shift toward hibernation and move to single or shared burrows where they'll spend the winter. The sharing can be either same-sex or opposite-sex pairs or groups. Juvenile tortoises tend to hibernate in their own, smaller burrows.

The burrow system, then, serves as a matrix that structures *Gopherus* populations in ways that change seasonally. The patterns of burrow use indicate a social network that cannot be entirely explained by random associations, and seasonal burrow switching may help secure food and mating opportunities. Tortoises might not form the intense social bonds of mammals, but some social ordering is apparent. Stressful events in these desert tortoise populations may disrupt the social structure; Mojave Desert Tortoises that have been translocated don't switch burrows as often. The importance of the burrow network merits attention in the management of Mojave Desert Tortoises and other burrowing tortoise populations (Chap. 10).

Migrations

All tortoises show high fidelity to their home ranges, making daily or seasonal movements within a specific, limited area. Gopher Tortoises in southern Florida, for example, remain in flatwood habitats even during seasonal flooding rains. They avoid high waters by moving to higher ground within their home ranges. When tortoises make longer movements, those behaviors are often tied to environmental conditions. As ectotherms ("cold-blooded" animals), tortoises are more affected by environmental variation than endotherms ("warm-blooded" animals). Thus movements at night, when temperatures are cooler, are infrequent. An exception is the Bolson Tortoise, which shifts to nocturnal foraging during the height of summer.

There is some evidence, however, that tortoises make longer, one-way journeys to escape unsuitable conditions or perhaps emigrate from crowded habitats. Observing one-way tortoise migrations in the wild is difficult. But there are anecdotal accounts from the 1930s and 1940s of biologists who reported what appeared to be migrations of Berlandier's Tortoises, possibly because of flooding. The fact that tortoises have managed to colonize a range of habitats on five continents is testimony to their migratory abilities (Chap. 8).

Populations

The actual size or number of living individuals within a species is unknown for all tortoises. Even for the many species whose populations are clinging to the edge of survival, we don't know the precise number of surviving individuals because one or two individuals, perhaps more, are likely to be missed even in the most thorough search. Few species live in a continuous and sufficiently small geographic area to permit a complete census. The population sizes reported here are estimates.

Estimating Abundance

Even the population size of the Aldabra Giant Tortoises is an estimate. Despite the limited land area of ~168 km^2 (roughly half of which is uninhabitable), not every tortoise can be found and counted. The various estimates of 129,000 individuals in 1973–1974 and ~100,000 in 1997 for the total Aldabra population were derived from mark and recapture estimates of the population. To obtain these estimates, individual tortoises are uniquely marked and counted at the beginning of the study, then several months or a year later, the population is counted again. Based on the proportion of marked to unmarked, the total number of individuals in the area is estimated.

Censusing across the numerous islands and their populations yielded a total estimate for Galápagos Tortoises of ~18,000 individuals in 2020. The contrasting population sizes between Aldabra and Galápagos likely represent different levels of human disturbance, as well as the time gap between measurements. Total population estimates for other tortoise species are less readily available and vary as tortoise populations continue to decline due to human harvesting, habitat modifications, and climate changes (Chap. 9). For example, the once widespread northwestern African populations of Spur-thighed Tortoises experienced unsustainable harvesting (~100,000 individuals annually) during the 1970s as well as human conversion of its habitat, leading to its remnant populations today.

Tortoise Density

Density, or the number of individuals of a species living in a specified area, is typically calculated for tortoises as the number of individuals (both juveniles and adults) present in a hectare (10,000 m^2, roughly the size of five football fields) of habitat. Known densities of tortoise populations worldwide range from super low (e.g., 0.0021 African Spurred Tortoises/ha in Burkina Faso) to orders of magnitude higher in other species (up to 57 individuals/ha). Where and when density is measured is a large determinant of the numbers. For example, for Spur-thighed Tortoises in Morocco, there were 3–5 individuals/ha in the 1970s; by the early 2000s, that number had diminished to 0–0.5 individuals/ha.

The density of tortoises mirrors their habitat quality and consequent availability of food, which shapes reproductive rates. Even tortoise species that are widespread may occur in low densities, such as Leopard Tortoises in Ethiopia with an estimated density of 0.03 individuals/ha or lower. Within a species, density differs based on habit. For example, Aldabra Tortoises occupy two atoll islands (Malabar and Grande Terre) with distinct habitats. On Malabar, where the tortoise population is confined to a narrow seaside border of open woodland, the density in 1997 was about 12 individuals/ha. In contrast, on Grande Terre, where tortoises live in grasslands bordered by a scattered scrub forest, their density was 18 individuals/ha.

These differences may also reflect past survivorship trends. About 20 years earlier, in 1974, the density of tortoises on Grande Terre was assessed at ~27 individuals/ha. When severe drought reduced the seasonal regrowth of grassland, the Grand Terre population faced starvation and declined. The Malabar population density, in contrast, was only about 5 individuals/ha in 1974. The 20-year increase in density of tortoises on Malabar likely reflected their slow, continual recovery from past human harvest that ended with the establishment of a biological research station in Aldabra.

Population Structure

Genetic population structure can be used to make inferences about how animals have dispersed and the resultant gene flow (movement of genes within and out of a population). Mojave Desert Tortoises, for example, can be divided into a set (of four to eight, depending on analyses) of distinct genetic clusters along a north-south gradient. Tortoises within a 200- to 276-km straight line radius from each tend to be genetically related enough to be considered a single genetic unit. The geographic locations of the clusters shows that long-term movements of tortoises have been restricted by landscape features like mountains, rock formations, and vegetation that are hard to traverse.

Similarly, subpopulations of Aldabra Giant Tortoises occur because of their inability to cross jagged, rocky coastal areas that lack shade. The evolution of population structure on islands results from geologic and climatic events that produce habitat differences, as well as physical discontinuity of insular populations by water such as those on Malabar and Grande Terre. If individuals from separate islands have not mated with each over a long period, you'd expect the populations to evolve independently and gradually diverge. But the mitochondrial DNA (mtDNA) among Aldabra Tortoise populations is uniform across islands, despite there being little or no reproductive exchange. A recent genetic study of nuclear DNA showed significant population structuring, illuminating at least some divergence of these populations because of the ecological and physical barriers.

In the case of Galápagos Tortoises, the genetic population structure is complex and still resolving scientifically. A recent mitochondrial genome study reveals what appears to be an additional lineage of giant tortoise from cave remains on Isla San Cristóbal, one that is distinct from the San Cristóbal Giant Tortoise (*C. n. chatamensis*) living on the island today. San Cristóbal is pegged as one of the first islands colonized by a tortoise(s) that had floated from mainland South America (Chap. 8). How tortoises then radiated across the islands to ultimately comprise the 14 known subspecies (15 including the possible cave lineage) is still speculative, pending more genetic information from museum specimens and living tortoises.

Changing Densities

Africa has the greatest diversity of tortoise species as well as the broadest range of species sizes, from the near-giant African Spurred Tortoise to several dwarf tortoise species in southern Africa. Most African Spurred Tortoise populations have been eradicated or greatly reduced. The next largest species—the Leopard Tortoise—has the widest distribution, ranging throughout much of sub-Saharan Africa. Conservationists suggest that it's not endangered, given its wide distribution, yet population estimates yield densities of less than one tortoise per hectare, for example, 0.02 individuals/ha for farmland in contrast to 0.8 individuals/ha in a national park. For Home's Hinge-back, the outlook is not much better. In unprotected areas, tortoise densities are 0.2–0.9 individuals/ha, in contrast to 1.6 to 2.9 individuals/ha in protected areas.

In North America, various *Gopherus* species have been the subject of intense study on protected lands. Gopher Tortoises in southeastern North America show a range of population densities. In 2014, an estimate proposed ~1,000,000 tortoise in Florida, with fewer individuals in the outer edges of their distribution. In suitable habitat, their densities typically range from 10 to 20 individuals/km^2. Densities of Texas Gopher Tortoises are orders of magnitude higher in their preferred habitats, with estimates of 1,500–1,600 individuals/km^2. But tortoise densities continue to decline; for example, estimates of 2–18 Mojave Desert Tortoises per square kilometer in 2011 contrasts with an estimate of 3.9 individuals/km^2 for the entire species in 2019.

Populations reduced to unnaturally low densities may reach tipping points where there are not enough reproductively fit adults to replace individuals that die. Conditions of extreme heat and drought brought on by climate change may be a lethal combination for already-shrinking tortoise populations (Chap. 9).

Age-Sex Structure

Tortoise populations are discontinuous because the habitats they prefer tend to be patchy; for example, they are interrupted by geologic features such as rivers, have varying soil conditions that produce different plant communities, or are characterized by elevational changes that affect local climates. Within this mosaic of different micro-landscapes, tortoise populations have different demographics, such as number of individuals per unit area (density), ratio of immature individuals to adults, number of females to males, and differential growth rates of juveniles, females, and males. Like populations of other animals, tortoise populations are molded by the habitats they must occupy to survive and successfully reproduce. Their population structures are intimately correlated with habitat attributes.

For example, Hermann's Tortoises along the north coast of the Mediterranean Sea from eastern France to the Levant have existed alongside humans for millennia and currently populate a range of habitats, from highly disturbed to less disturbed. Comparisons of relatively undisturbed populations (forested slope of Massif des Maures in southeastern France) and disturbed populations (seaside in northern Greece) revealed contrasting population characteristics. The less disturbed French woodland had a density of just 1 adult/ha, with equal numbers of adult females and males (45% each). The Greek grassy heath had 45 adults/ha, with nearly double the number of males (at 56%) to females (31%). Juveniles were present in small (10% France and 13% Greece) though significant numbers.

A similar comparison is available from censuses of Sonoran Desert Tortoises populations. In most populations, sex ratios are equal or nearly so, and depending on the site, females may be equal sized or males somewhat larger (Fig. 6.7; Chap. 4). The greatest differences between populations are usually the number of juveniles, which are indicators of the reproductive recruitment and health of a population. Because of the relatively long life span (longevity) of tortoises, an unhealthy population can persist for a long time—likely more than a decade—but it is in constant decline as adults die.

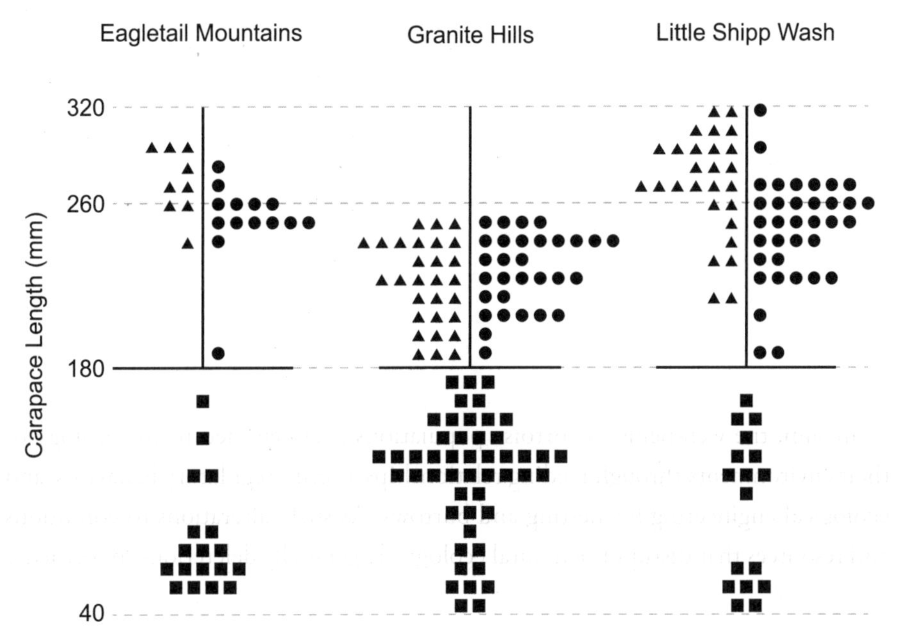

Figure 6.7
Size distribution (straight carapace length, in millimeters) comparison of three populations of Sonoran Desert Tortoises (*Gopherus morafkai*) in Arizona. Granite Hills lacks older, larger adults and has the strongest recruitment; Little Shipp has the broadest range of adult sizes and only modest recruitment; and the Eagletail Mountains population appears to have suffered a loss of adults, but its recruitment profile, although small, suggests that it will endure. Each row represents a 10-mm SCL size class. Adult males, triangles; adult females, circles; immatures, squares. Data from Averill-Murray et al. (2014, Fig. 6.2)

Population Regulation

The greatest mortality in all tortoise populations is of the eggs and young (Chap. 5). Many studies report 100% mortality of juveniles for most egg clutches from egg deposition to one year of age. Obviously, total mortality cannot exist for every egg clutch, or there would be no recruitment and the population would die out.

Predation pressure influences population size and growth. The contrast between giant tortoise population sizes on Aldabra and the Galápagos Islands is largely explained by the amount of predation on these disparate islands. Rats certainly arrived with humans to both islands, but they seemingly found Aldabra less habitable, and their populations remained small and predation light. In the Galápagos, in contrast, rat populations boomed, and they became major predators of juveniles and hatchlings. On Isla Pinzón, the tortoise population suffered no recruitment for more than

20 years owing to rat predation on eggs and juveniles. Only with the extermination of the rats was the tortoise population able to begin a slow recovery.

Annual weather fluctuations also regulate populations by influencing the quality and abundance of food. As noted above, the Aldabra Giant Tortoise population of Grande Terre experienced a sharp decline (of more than 20%) during a decade of drought that reduced grasses and increased the tortoises' exposure to sunlight as they sought forage further afield. Similar drought conditions have affected the desert tortoises of the southwestern United States and are implicated—along with the respiratory disease pandemic—in bringing the Mojave Desert Tortoise to the brink of extinction (Chap. 9).

In sum, the well-being of tortoise populations is closely tied to the ecology of their environments through feeding relationships, thermoregulatory behaviors, and ecological engineering for nesting and burrows. As such, alterations to conditions and resources that disrupt the natural ecology are generally deleterious to tortoises.

Plate 1
A Ruddy Mongoose (*Urva smithii*) biting an Indian Star Tortoise (*Geochelone elegans*). The tortoise is a mouthful, perhaps too large for the mongoose to crack its shell. Photograph by Sanka Karunaratne, Wikimedia CC-BY-SA-4.0

Plate 2
A Chaco Tortoise (*Chelonoidis chilensis*) walking. Its body is held off the ground and supported by three limbs throughout most of the walking cycle. Photograph by Walter S. Prado, Wikimedia CC-0-1.0

Plate 3
A male Asian Giant Tortoise (*Manouria emys*) flipped over, perhaps during a courtship battle. By waving its limbs, it can slowly shift onto its side and may be able to right itself. Photograph by Warren Garst, Wikimedia CC-BY-SA-4.0

Opposite

Plate 4
A saddle-backed Western Santa Cruz Giant Tortoise (*Chelonoidis n. porteri*). The flared front of the shell highlights its long, yellowish neck that allows it to browse rather than graze like its domed cousins. Photograph by Ken Dodd

Plate 5
A Sonoran Desert Tortoise (*Gopherus morafkai*) eating the red fruits ("tunas") of Prickly Pear (*Opuntia*) cacti. These fruits and pads are especially attractive and nutritious for the tortoises of western North America. Photograph from the National Park Service

Opposite

Plate 6

A Mojave Desert Tortoise (*Gopherus agassizii*) scraping rocks and caliche (calcium carbonate deposits) with its beak and tongue. Rocks provide salts and minerals, particularly calcium, for tortoises. Photograph by Daniel Elsbrock, National Park Service

Plate 7

An Egyptian Tortoise (*Testudo kleinmanni*) traversing its incredibly arid, sunny, sandy habitat in the Omayed Protected Area in western Egypt. Photograph by Hatem Moushir, Wikimedia CC-BY-SA-4.0

Below

Plate 8

An Aldabra Giant Tortoise (*Aldabrachelys gigantea*) drinking from a shallow coral-stone puddle filled by a predawn rain shower. These tiny, fleeting basins are the only source of fresh water near its grazing area. Photograph by G.R.Z.

Opposite

Plate 9
Early-morning grazing herd of Aldabra Giant Tortoises (*Aldabrachelys gigantea*) on tortoise turf of the Cinq Cases area of Aldabra. The dead tortoise in the bare sand foreground likely misjudged its internal heating rate and died from overheating. Photograph by J. C. Schaffer

Plate 10
Aldabra Giant Tortoises (*Aldabrachelys gigantea*) wedged beneath the few available trees. Taking advantage of the scarce shade prevents overheating and death. Photograph by J. C. Schaffer

Below

Plate 11
A numbered juvenile Mojave Desert Tortoise (*Gopherus agassizi*) from the Edwards Air Force Base Head Start Program eating thistle. Photograph courtesy of Edwards Air Force Base

Plate 12

Darwin's Finch removing insects from the skin of a Volcán Alcedo Giant Tortoise's (*Chelonoidis n. vandenburghi*) right foreleg. The tortoise's high-stance posture allows the finch to access all its skin. Photograph by Ken Dodd

Opposite

Plate 13

Tent Tortoises (*Psammobates tentorius*) mating. Two males followed the female to this spot of shade to safely court her. The slightly larger male was successful this time. Photograph by Thomas Leuteritz

Plate 14

Male vocalizing while this pair of Hermann's Tortoises (*Testudo hermanni*) copulate. Noisy mating behavior is common in species of *Testudo* and can be heard from many meters away. Photograph by Richard Mayer, Wikimedia CC-BY-SA-3.0

Plate 15
A Leopard Tortoise (*Stigmochelys pardalis*) laying its eggs in a large nest chamber. Because of the pliable sandy soil, the chamber is much larger than a nest made in compacted soil. Photograph by Louisvdw, Wikimedia

Plate 16
The Leopard Tortoise starting to close its nest cavity. The darkened soil is wet from its bladder fluid. Adding moisture softens the soil for digging and then, at this stage, prevents the chamber from collapsing. Photograph by Louisvdw, Wikimedia

Plate 17
Coconut crabs (*Birgus latro*) trying to prey on eggs of a nesting Aldabra Giant Tortoise (*Aldabrachelys gigantea*), who has lowered the rear of her shell to protect the eggs as she lays them. Photograph by Ian Swingland

Plate 18
Predation of a juvenile Aldabra Giant Tortoise (*Aldabrachelys gigantea*) by an endemic White-throated Rail (*Dryolimnas cuvieri*) poking its beak through the plastron. Photograph by Ian Swingland

Opposite

Plate 19
Aldabra Giant Tortoises (*Aldabrachelys gigantea*) trapped on a tidal flat by a quickly rising tide. If a tortoise does not have a foothold, the falling tide will carry it into the lagoon. Occasionally, one gets caught in the lagoon's exit flow and washes out to sea. Photograph by Ian Swingland

Plate 20
Adult female Aldabra Giant Tortoise (*Aldabrachelys gigantea*) that walked ashore on a beach in Tanzania in 2004 after floating ~740 km from Aldabra. The size of her barnacles suggests a float time of at least six weeks. Photograph by Catherine Joynson-Hicks

Opposite

Plate 21

Travancore Tortoise (*Indotestudo travancorica*) seen active at night in a wet tropical forest. Most tortoise species are strictly diurnal. Photograph by Swati Sidhu, Wikimedia CC-BY-SA-4.0

Plate 22

Gopher Tortoise (*Gopherus polyphemus*) enlarging its burrow. When digging a burrow, tortoises sweep their front legs backward to throw sand or soil away from the hole. Photograph by Andrea Westmoreland, Wikimedia CC-BY-SA-2.0

Below

Plate 23

Well-trodden path on Aldabra. Both Aldabra and Galápagos Giant Tortoises scour pathways through their habitats. This Aldabra Tortoise (*Aldabrachelys gigantea*) bears a titanium disk for long-term identification in mark and recapture studies of population dynamics. Photograph by Ian Swingland

Plate 24
Ploughshare Tortoises (*Astrochelys yniphora* or "Angonoka") in the Ampijoroa breeding facility of the Durrell Wildlife Conservation Trust. Photograph by Ken Dodd

Plate 25
Yellow-footed Tortoise (*Chelonoidis denticulatus*) near a buttress tree in lush tropical rainforest habitat of Yasuní National Park, Ecuador. Photograph by Geoff Gallice, Wikimedia, CC-BY-2.0

CHAPTER 7

Today's Species

Tortoise Diversity

Origins of Tortoise Diversity

The species diversity of turtles, including tortoises, is lower than that of most other living reptiles, mammals, or birds. The suite of 357 turtle species known to be living today is sparse compared to the thousands of species in those other groups, especially given turtles are an ancient group of organisms that have demonstrated resilience to major changes over their more than 220-million-year history on Earth (Chaps. 3 and 8).

Turtles arose in the Early Triassic and diversified into more species as their distribution expanded worldwide. They also radiated into higher latitudes during the warming periods of the Jurassic and Cretaceous. Their diversification appears to have plateaued by the end of the Cretaceous (66 million years ago when the large dinosaurs became extinct). Some turtles survived the mass extinction event, then again diversified in the Late Paleocene. One of those diversifications resulted in the lineage that would become tortoises. Tortoises continued diversifying through the Eocene and on into the Pleistocene ice ages. With the appearance of humans, extinctions began. On continental mainlands, large tortoise species disappeared, apparently due to predation by humans.

Yet despite these bursts of diversification, the total biodiversity of turtles has been limited by high turnover, with many turtle species going extinct. Since humans multiplied and spread over the globe, the diversity of tortoises and other turtles has plummeted as habitats are eliminated, turtles are harvested, and predators are introduced. Today, the highest species diversity among tortoises continues to be in the original epicenter of their diversification in sub-Saharan Africa and islands off its coast, but most tortoise species are threatened with extinction.

Classification of Tortoises

All pre-tortoise species and genera are classified as Pan-Testudinidae because they lack some characteristics of modern tortoises, the Testudinidae. Within the modern tortoises are three lineages whose ancestry extends back to the mid-Eocene, about 40 to 42 mya or earlier (Chap. 8). The group of tortoises known to have the earliest origins is the Asian *Manouria* (Manouriinae). The *Gopherus* lineage (Xerobatiinae) in North America is nearly as old. Each lineage is recognized as a separate subfamily within the Testudinidae.

All other tortoises are classified in a third subfamily, Testudininae, which includes 39 species of living (extant) tortoises and a nearly global fossil record that includes some extinct species. Even within this subfamily, there are distinct evolutionary lines recognized as the Testudona and the Geochelona. The Testudona group contains *Malacochersus*, *Indotestudo*, and the *Testudo* clade. Even though these generic lineages are related, they diverged in the early Late Eocene. All Testudona are medium to small tortoises, none exceeding 55 cm straight carapace length (SCL). The Geochelona contains all the remaining living genera and several extinct genera, from the modern-day dwarfs *Chersobius* to the extinct *Titanochelon*, the real giants of the tortoise world.

Diversity of Today's Tortoises across Continents

In addition to all continents except Antarctica and Australia, tortoises occur, or did occur, on some distant oceanic islands: Mascarenes, Seychelles, and Galápagos (Each continent has its own unique suite of tortoises. For Europe, southwest Asia, and northern Africa, there's a single genus, *Testudo*, composed of five species and more than a dozen subspecies or varieties. Sub-Saharan Africa has a rich tortoise fauna that includes 21 species across nine genera. Madagascar and the Indian Ocean islands have experienced particularly high tortoise extinction since humans arrived on the scene. Of the four genera and 11 species that once existed in the Indian Ocean region, there are three genera with five species remaining. Tropical Asia harbors three genera of tortoises comprising seven species. All six species of North American tortoises are in the genus *Gopherus*. South America has three species of tortoise, all in the genus *Chelonoidis*. The Galápagos Islands historically were the habitat for 15 distinct populations (each now designated as a distinct subspecies) of *Chelonoidis niger*, of which 12 populations remain today.

Eurasian Tortoises

A group of five small- to medium-sized tortoises in the genus *Testudo* live in the grasslands and open forests from the Atlantic coast of north Africa to the grassy steppes of southwest Asia (Fig. 7.1). The Spur-thighed Tortoise is the most widespread of the five and likely the most abundant one as well. Certainly, it and Hermann's Tortoise are the most familiar tortoises to Europeans who for centuries kept them as garden pets. Their capture and keeping were made illegal in the late 1970s as declining populations and local extirpations became evident throughout their Mediterranean distribution. Two of the five species have maximum adult carapace lengths that classify them as dwarfs, that is, less than 20 cm SCL.

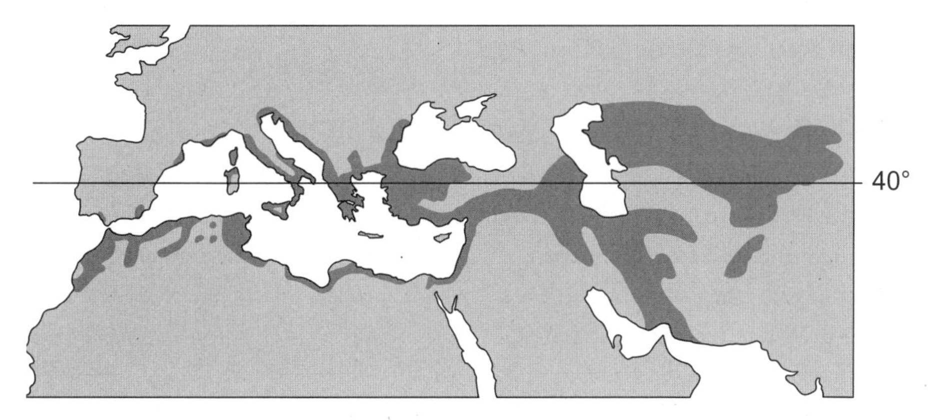

Figure 7.1
Generalized distribution of extant Eurasian tortoises; all species are currently considered members of the genus *Testudo*. Data from the distributions associated with species accounts in Turtle Taxonomy Working Group (2021)

Spur-thighed Tortoise (*Testudo graeca*)

♀♀ 8–34 cm; ♂♂ 9–27 cm adult SCL, maximum reported size to 46 cm

A complex of 10 small- to medium-sized tortoise subspecies occurs in coastal grassland and open woodland habitats of northern Africa (Morocco to western Libya) and in the northeastern Mediterranean from Greece into Turkey and Levant eastward to Iran (Fig. 2.2*B*). All have domed carapaces and a transverse plastral hinge between the rearmost pair of plastral bones. The back of each thigh has a cluster of small keratinous spines surrounding a larger spine. Although the function of the spurs is unknown, they likely help protect the soft tissues around the tail.

Hermann's Tortoise (*Testudo hermanni*)

♀♀ 13–25 cm; ♂♂ 12–24 cm adult SCL

This moderately small species (Fig. 2.2*A*, Plate 14) with a domed carapace and flared posterior marginals lives in a mosaic of habitats from olive tree groves and other open forests. They are most common in the natural scrubby marquis of the north Mediterranean coast from eastern Spain to the Balkans, including Corsica, peninsular Italy, Mallorca, Sardinia, and Sicily, with a hot spot of genetic diversity in the Calabria region of southern Italy. In most localities, males are significantly smaller than females. Further, adult size is geographically variable, likely owing to variations in local climate and availability of local food resources. Sometimes this species is recognized a distinct genus *Agrionemys*.

Steppe Tortoise (*Testudo horsfieldii*)

♀♀ 9–19 cm; ♂♂ 9–15 cm adult SCL

This small species with a domed carapace occurs in the steppe grassland of south-central Asia from the Caspian Sea to the Hindu Kush. Commonly called the "Russian Tortoise," these tortoises are popular in the pet trade and vulnerable in their native range where severe habitat fragmentation has occurred. An estimated 100,000 individuals are taken from the wild per year in Uzbekistan alone. There are five subspecies of Steppe Tortoise.

Egyptian Tortoise (*Testudo kleinmanni*)

♀♀ usually to 13.1 cm; 10.6 mm ♂♂ mean adult SCL; largest reported ♀ 14.4 cm SCL

This tiny, dome-shelled species (Fig. 10.1, Plate 7) occurs in the desert and semidesert habitats of the highly seasonal Mediterranean coasts of eastern Libya through Egypt and on into Israel. Its habitat ranges from sandy, gravel plains and adjacent salt marsh habitats to Mediterranean scrub. Most individuals have two distinctive brown or black triangular markings pointing toward their rear end on an otherwise yellowish plastron. Owing to habitat loss and overharvest for the pet trade, this species is critically endangered and effectively extinct in the wild in Egypt.

Marginated Tortoise (*Testudo marginata*)

♀♀ 22–34 cm; ♂♂ 22–34 cm adult SCL

This moderate-sized species frequents dense, thorny scrub around rocky outcrops, but it's also found in other habitats including coastal dunes and agricultural land, such as olive groves. It ranges throughout most of Greece and into Albania. The rear marginals of its domed carapace usually flare upward at the back like the edge of a bell. Many individuals bear scarred carapaces owing to frequent human-caused fires in its primary grassland-scrub habitat.

Sub-Saharan Tortoises

Sub-Saharan Africa has the greatest diversity of tortoises, with 39% of living species (Fig. 7.2). One species reaches giant proportions (*Centrochelys sulcata*), and several species are the world's smallest tortoises ("dwarfs"; *Chersobius* spp.). Sub-Saharan Africa, along with Madagascar, is also home to the only species that have a moveable carapace hinge—the "Hinge-backs"—a feature that comes into play during egg laying (Chap. 4). Tortoise species occur throughout the African continent except within parts of the Congo Basin, perhaps owing to its dense forests and the rarity of ground-level vegetation and invertebrates for them to eat.

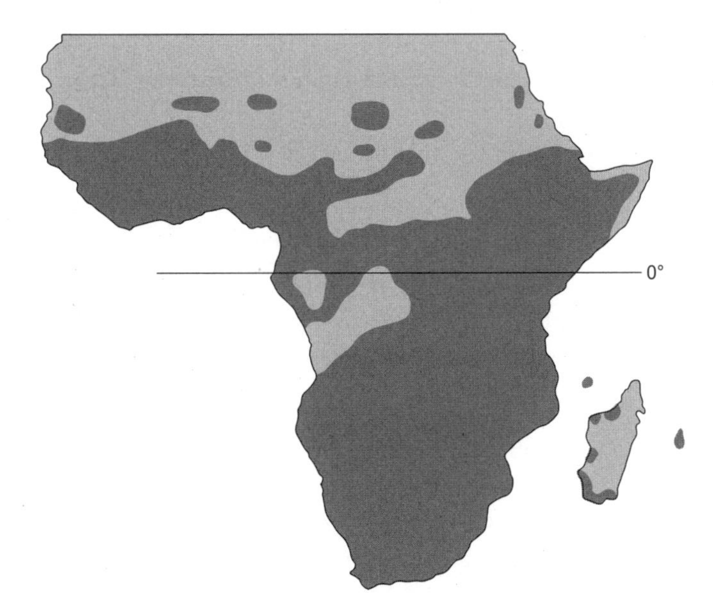

Figure 7.2
Generalized distribution of all extant species of tortoises of sub-Saharan Africa and the western Indian Ocean. African tortoises consist of 8 genera and 20 species. Data from Turtle Taxonomy Working Group (2021)

African Spurred Tortoise (*Centrochelys sulcata*)

♀♀ 34–64 cm; ♂♂ 38–97 cm adult SCL

Populations of this species usually occur near seasonally flowing streams and rivers, with a preference for stabilized dunes that provide abundant annual plants during the wet season and a substrate for burrowing. Despite their massive size, they dig long burrows as means to avoid the extreme arid conditions and high temperatures of their habitat. In contrast to their more rectangular-shaped burrowing cousins (*Gopherus*) in North America, African Spurred carapaces are dome shaped.

Angulate Tortoise (*Chersina angulata*)

♀♀ to 17 cm; ♂♂ to 19 cm mean adult SCL; an exceptionally large captive male is 35 cm

This moderately small species has a domed, rectangular carapace (Fig. 2.4). Angulate Tortoises inhabit the scrublands of coastal plains and escarpment of southern Namibia and South Africa to Elizabethtown, as well as several offshore islands. Though they favor sandy substrates, they are also found in rocky areas. Females have an unusual reproductive pattern for a Mediterranean climate tortoise. Unlike European tortoises that nest seasonally in spring and summer, Angulate Tortoise females may nest year-round with just a brief summer hiatus. They lay one egg at a time, a characteristic otherwise seen only in subtropical and tropical species.

Karoo Dwarf Tortoise (*Chersobius boulengeri*)

♀♀ 6–16 cm; ♂♂ 5–13 cm adult SCL

This species is one of the world's smallest. Like all *Chersobius*, it has five claws on each foot. Its shell is rectangular and flat. It lives amid rocky ridges and outcrops ("koppies") in the succulent veld and desert grasslands of southern and south-central South Africa. These tortoises spend 80%–90% of their time sheltering in the rocks where available and are thus difficult to study. They forage during an hour or two in the afternoon and evening, reflecting low resource demands. They can regulate their body temperatures with little basking outside of their rocky retreats. Their secretive daily habits may relate to their small body size, which makes them particularly vulnerable to predators, including crows.

Speckled Dwarf Tortoise (*Chersobius signatus*)

♀♀ 5–11 cm; ♂♂ 6–10 cm adult SCL

This species is the world's smallest tortoise. Males are smaller than females. Its carapace has a speckled pattern, and like its relatives (*C. boulengeri* and *C. solus*), it has a rectangular flat carapace with a rough appearance owing to each vertebral, costal, and marginal scute bearing a low pyramidal stack of the previous years' scutes. It lives in west coastal South Africa in rocky veld landscapes, where rock crevices shelter it from overheating during the summer. Speckled tortoises appear to become sexually mature in about two years, which is the fastest known maturity of any tortoise.

Nama Dwarf Tortoise or Nama Padloper (*Chersobius solus*)

♀♀ 8–14 cm; ♂♂ 8–14 cm adult SCL

This small species with a rectangular, flat, golden-brown carapace occurs in a limited area of the semidesert grassy scrubland of southwestern Namibia. These tortoises frequent rocky areas and regularly shelter beneath west-facing rock slabs. Their diminutive size and reduced bony carapaces make them highly susceptible to predation, and likely explain their small home ranges and cautious foraging behavior. Their beaks end in three points (tricuspid), a feature that distinguishes them from their closest relatives, which tend to have either two points (*C. signatus*) or none (*C. boulengeri*).

Parrot-beaked Tortoise or Common Padloper (*Homopus areolatus*)

♀♀ and ♂♂ to 10–13 cm adult SCL, although ♀♀ occasionally to 16.8 cm

This small species (Fig. 2.6*C*) lives in the South African coast veld from Cape Town to East London. Compared to the other species in the genus, this tortoise favors the less arid habitats, such as the evergreen shrubland ("fynbos") in the Eastern Cape. Like its conspecific relative *H. femoralis*, Common Padlopers have domed rectangular carapaces and low-tacked dorsal scutes. They are sexually dimorphic, with red carapaces in males and green ones in females.

Greater Dwarf Tortoise or Greater Padloper (*Homopus femoralis*)

♀♀ 10–17 cm; ♂♂ 10–13 cm adult SCL

This small species lives in the grasslands and bushlands of central South Africa to elevations of 1900 m above sea level. Compared to the other species in the genus *Homopus*, it's an inland tortoise, restricted to habitats with higher rainfall. Greater Dwarfs have the broadest distribution of any padloper, yet their biology is poorly known. They are typically found under vegetation, with their activity seemingly triggered by rain. Climate change is expected to change the grassland habitat on which they depend.

Bell's Hinge-back Tortoise (*Kinixys belliana*)

♀♀ 14–22 cm; ♂♂ 14–22 cm adult SCL, not sexually dimorphic

This small species occupies savannah and open dry woodlands of central Africa diagonally from Angola to Sudan and Ethiopia. These Hinge-backs prefer dense patches of vegetation, often near bodies of water. Bell's Hinge-back and its relatives (Speke's Hinge-back, the Leopard Tortoise, and African Spurred Tortoise) share the broadest distributions of all African tortoises, each spanning multiple countries and climatic regimes.

Forest Hinge-back Tortoise (*Kinixys erosa*)

♀♀ to 30 cm, ♂♂ to 40 cm adult SCL; plastral lengths ♀♀ 17–21 cm; ♂♂ 18–23 cm

This small- to medium-sized species (Fig 4.6*B)* lives in a variety of closed canopy forest habitats in central to coastal West Africa (Sierra Leone to the Congo). Like other *Kinixys*, these tortoises are strongly omnivorous, eating mushrooms, seeds, and invertebrates, likely compensating for the paucity of edible vegetation at ground level.

Home's Hinge-back Tortoise (*Kinixys homeana*)

♀♀ 15–22 cm; ♂♂ 11–22 cm adult SCL; adult female typically larger than males

This small species (Fig 4.6*A*) lives in a variety of forests, including rainforest. These tortoises prefer forest gaps and swampy areas along the coast of the African Gulf of Guinea, although there are isolated interior populations (likely introduced) in drier forests in the Republic of Central Africa. Home's Hinge-backs become especially active after rainfall and have been observed dispersing by floating on water.

Lobatse Hinge-back Tortoise (*Kinixys lobatsiana*)

♀♀ to 20 cm; ♂♂ to 17 cm adult SCL

This moderate-sized species inhabits grasslands and scrub of northern South Africa, edging into Botswana. It favors rocky hillsides where cover is available. This species appears to be the most arid-adapted Hinge-back and is most active during the brief summer rainy season, spending the rest of the year in rock crevices or animal burrows.

Natal Hinge-back Tortoise (*Kinixys natalensis*)

♀♀ to 16 cm; ♂♂ to 13 cm adult SCL

This small species occurs in areas of rocky outcrops and ridges in thornbush-dominated "bushveld" at several thousand feet of elevation from eastern South Africa edging into southern Mozambique along the Lebombo Mountains. They prefer dry conditions and are thus absent from coastal and forest habitats. Yet they tend to be most active after rainfall. They are the only *Kinixys* with a three-pointed (tricuspid) beak.

Western Hinge-back Tortoise (*Kinixys nogueyi*)

♀♀ to 20 cm; ♂♂ to 17 cm adult SCL

This small species lives in wet savannas, rainforest patches, and hilly forests of the Gulf of Guinea area from Guinea–Sierra Leone to central Africa. This is the only Hinge-back with four claws on its front feet. Like other *Kinixys* species, its diet is omnivorous. It transitions from the animal food it favors in the dry season to a predominantly vegetarian diet in the rainy season.

Speke's Hinge-back Tortoise (*Kinixys spekii*)

♀♀ 15–22 cm; ♂♂ to 17 cm adult SCL

This small species has a broad inland distribution in sub-Saharan Africa, occupying most of inland Democratic Republic of Congo and extending into 11 other surrounding countries. It has a wide habitat tolerance that includes tropical savannas as well as forests. Speke's Hinge-backs are preyed on by mammals, which may remove part of the plastron, and by African hornbills, which punch holes through the shell. Its high mortality rates are offset by high juvenile growth rates.

Southeastern Hinge-back Tortoise (*Kinixys zombensis*)

♀♀ and ♂♂ 17–22 cm adult SCL

This moderate-sized species lives in coastal habitats from southern Kenya and Tanzania to extreme northeastern South Africa. This species lives mainly in coastal grasslands, savannah woodlands, and dune forests. They have also been observed in man-made habitats such as plantations and gardens. Like box turtles, they take refuge in depressions under leaf litter. Populations of Southeastern Hinge-backs also inhabit an area of just a handful of square kilometers in Madagascar; these populations were introduced by humans within the past 1,000 years but now constitute a subspecies (*K. z. domerguei*).

Pancake Tortoise (*Malacochersus tornieri*)

♀♀ 13–18 cm; ♂♂ 13–17 cm adult SCL

This species of small tortoise (Figs. 2.6*E*, 7.3) has a flat, broad carapace, a unique shape for a tortoise. The anterior and posterior openings of the shell are also large relative to the size of the tortoise, which allows for greater leg movement (Chap. 2).

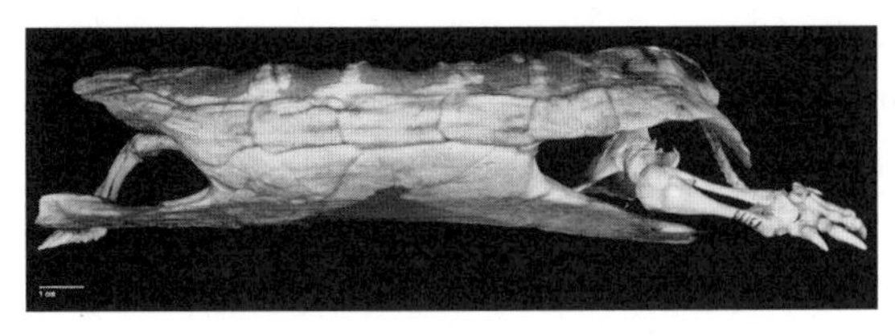

A

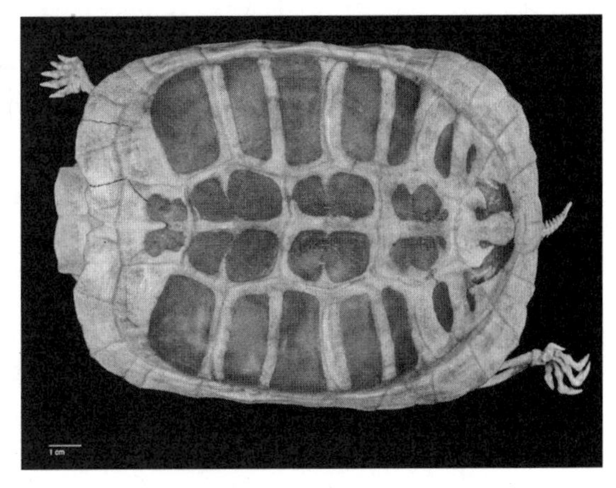

B

Figure 7.3
Carapace skeleton of a Pancake Tortoise (*Malacochersus tornieri*) without its scutes. Note the extensive areas of cartilage that lighten the weight of the shell and give it increased flexibility. (A) Side view. (B) Top view. US National Museum, USNM 73198

These tortoises move quickly (nearly running) to escape into the rocky outcrops ("koppies") amid the grasslands of Kenya, Tanzania, and northern Zambia. Its flattened and flexible carapace allows a Pancake Tortoise to wedge itself into rock crevices and avoid being extracted by predators.

Geometric Tortoise (*Psammobates geometricus*)

♀♀ 12–16 cm; ♂♂ 10–12 cm adult SCL

This small species with a domed shell lives in the shrublands of the southwest Cape area of South Africa. The females, which are larger than males, have bigger home ranges, perhaps because of reproductive nutrition needs. Only during the wet nesting season when food resources are concentrated do females stay more localized. Males increase their home ranges during the non-nesting season, when they may be in search of mating opportunities.

Serrated Tent Tortoise (*Psammobates oculifer*)

♀♀ 8–15 cm; ♂♂ 10–12 cm adult SCL

This small, dome shelled species occurs broadly in the arid savannas and scrub deserts of central Namibia and central South Africa. They favor sandy substrates. The wider spacing of the hind feet and increased rear shell opening of males to females may give males extra mobility for contesting dominance, courting, and copulation. Serrated Tent Tortoises take refuge in tall grasses during cooler times of the year, and in short grasses, spiny shrubs, or mammal burrows when conditions are hot.

Tent Tortoise (*Psammobates tentorius*)

♀♀ 10–15 cm; ♂♂ 9–13 cm adult SCL

This small species (Fig. 6.1, Plate 13) lives in a variety of grass and bushland habitats of southern Namibia and the western half of South Africa. *P. tentorius* consists of three geographic races; all possess a radiating pattern on the vertebral and costal scutes. They are strongly herbivorous, feeding on grasses, succulents, herbs, and other plants. They tend to feed in the early morning or late afternoon, sheltering under vegetation for most of the day, a crepuscular pattern.

Leopard Tortoise (*Stigmochelys pardalis*)

♀♀ and ♂♂ typically to 30–50 cm adult SCL, maximum reported 72 mm SCL

This large species (Plates 15, 16) with a high, dome-shaped carapace is the most widespread of all African tortoises. It occurs in eastern Africa from Ethiopia and southern Sudan southward into south-central South Africa. Thus it uses the most habitat types of any sub-Saharan species, ranging from heathland vegetation in South Africa to grassland in the northeast, often associated with hilly/mountainous terrain. The Leopard Tortoise's size also varies geographically, with smaller individuals found in dryer savannah and larger individuals in wetter locations, such as woodlands.

Tortoises of Madagascar and Indian Ocean Islands

Madagascar and the islands of the western Indian Ocean have suffered the greatest loss of tortoise species globally. All giant tortoises in this region, except those of the Aldabra Atoll, were extinct by the beginning of the 20th century or earlier. Today, even the smaller species of Madagascar are threatened by local consumption and international trade. Only the continual efforts of conservation groups have prevented the extinction of the surviving species (Chap. 10).

Madagascar Giant Tortoise (*Aldabrachelys abrupta*)

Extinct

♀♀ and ♂♂ adult SCL to 113 cm

This species was large with a distinctly domed shell. It likely ranged widely on the western half of the island. Limited fossil evidence suggests high abundance before the arrival (ca. 100–200 CE) and settlement (~490 CE) of humans, who drove them to extinction by the 11th or 12th century. Although *A. grandieri* has been proposed as the source population for the Aldabra Giant Tortoises, new genetic evidence demonstrate that it and *Aldabrachelys abrupta* are distinct species.

Aldabra Giant Tortoise (*Aldabrachelys gigantea*)

♀♀ 61–79 cm; ♂♂ 59–114 cm adult SCL

This giant species occurs naturally on Aldabra Atoll, a raised, 34-km-long limestone reef with a maximum width of 14.5 km (Figs. 2.1*A*, 2.3, 4.5; Plates 8, 9, 10, 18, 19, 20, 23). Most individuals bear distinctly domed shells. Much of the atoll's islands are covered by a low coastal scrub forest, except for a large grassy plain along much of the southeastern coastline. These grasses are closely cropped into "tortoise turf" by a large population of tortoises. In the absence of the other Indian Ocean oceanic island tortoises (*Cylindraspis*, now extinct), much of our knowledge of giant tortoise biology derives from this species and the Galápagos giants, *Chelonoidis niger*.

Grandidier's Giant Tortoise (*Aldabrachelys grandidieri*)

♀♀ and ♂♂ to 125 cm adult SCL

This Malagasian species was large with a heavy, domed, slightly flattened shell. It likely occurred widely on the western half of the island, apparently with a broad distribution across habitats on the coast and the cooler highlands. Limited fossil evidence suggests a high abundance before the arrival of humans, who harvested them to extinction by the middle of the 13th century. These tortoises were grazers rather than browsers, potentially playing a similar ecological role to large grazing mammals like bison, including the dispersal of seeds from the Baobabs and other plants on Madagascar.

Seychelles Giant Tortoises (*Aldabrachelys gigantea*)

Extinct

Native tortoises lived on the granitic Seychelles before the arrival of humans. Genetic analysis of subfossil bones and the earliest known Seychelles' specimens in museum collections identify them as *Aldabrachelys gigantea*. Recent data suggest that an *Aldabrachelys* lineage related to *A. abrupta* colonized the Seychelles from Madagascar and became *A. gigantea* (Chap. 8). Once the Seychelles were colonized by humans, the native tortoises were rapidly harvested to extinction, by the middle of the 18th century if not earlier. The three species (*A. arnoldi*, *A. daudinii*, *A. hololissa*) proposed as native to the Seychelles prior to human arrival are variant morphological forms of *A. gigantea* and likely derived from Aldabra individuals; even before the extinction of the native Seychelles tortoises, the importation of Aldabra tortoises had begun.

Mascarene Giant Tortoises (*Cylindraspis*)

Extinct

Of the Mascarene Islands, three of them (Mauritius, Réunion, Rodriques) supported populations of giant tortoises (*Cylindrapis indica* to 110 cm plastron length [PL],

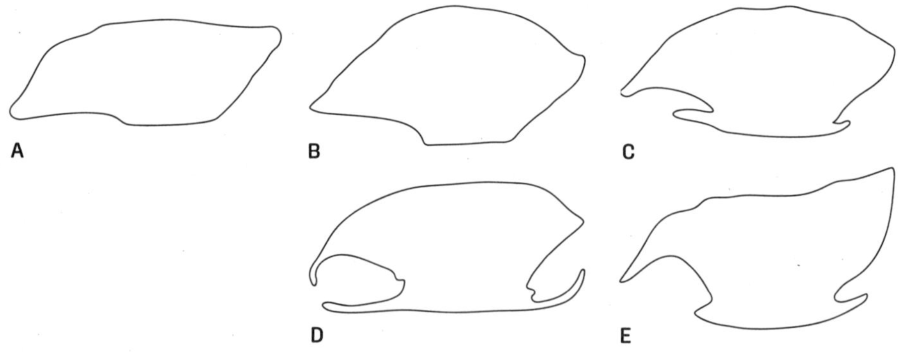

Figure 7.4

Shell shape in adult tortoises of the extinct Mascarene genus (*Cylindraspis*): **(A)** *C. indica*, rectangular elongate shell; **(B)** *C. inepta*, domed shell; **(C)** *C. peltastes*, domed shell; **(D)** *C. triserrata*; rectangular domed; **(E)** *C. vosmaeri*, saddleback shell. The front and rear limb openings are not well defined for *C. indica* and *C. inepta* because the fossil illustrations used did not show each plastron. Lateral profile outlines from photographs in the Cylindraspis species accounts in Turtle Taxonomy Working Group (2021)

Réunion; *C. inepta* to 104 cm PL, Mauritius; *C. peltastes* to 46 cm SCL, Rodriques; *C. triserriata* to 75 CL, Mauritius; *C. vosmaeri* to 109 cm SCL, Rodriques) when they were discovered by Europeans in the 1600s. A century later, these endemic tortoises (Fig. 7.4) were largely extinct due to human harvesting (Chap. 9).

Radiated Tortoise (*Astrochelys radiata*)

♀♀ 24–36 cm; ♂♂ 28–40 cm adult SCL

This moderate-sized, distinctly domed species of southern coastal Madagascar lives in a variety of habitats from coastal dune scrub to dry deciduous spiny forest and inland scrub forest on high plateaus. Female reproductive output is variable, with individuals nesting one to three times each year and laying up to five eggs per clutch. Nesting is triggered more by environmental conditions than courtship cues (Fig. 4.3*E–H*), typically starting with the first rains of the wet season. This reproductive plasticity may be critical for the survival of this imperiled species.

Bour's Tortoise (*Astrochelys rogerbouri*)

Estimated 50 cm SCL

This tortoise is recently (2023) described from a single subfossil tibia that was discovered at the same site as one of the subfossil specimens of *A. abrupta.*

Ploughshare Tortoise (*Astrochelys yniphora*)

♀♀ 31–43 cm; ♂♂ 36–49 cm adult SCL

This moderate-sized species (Fig. 4.1, Plate 24) lives in patches of palm savannah and thorn-scrub thickets in the Baly Baly region (~700 km^2) on the northwest coast of Madagascar. The massive and protruding gular scales of adult males are used in courtship battles and give these tortoises their common name. Threatened by collection for the pet trade and habitat alteration for grazing of Zebu cattle, Ploughshare Tortoises are the rarest of all living turtles. They survive only through the efforts of the Durrell Wildlife Conservation Trust and other conservation efforts.

Spider Tortoise (*Pyxis arachnoides*)

♀♀ 9–15 cm; ♂♂ 8–14 cm adult SCL

This small species is a dwarf tortoise that inhabits the dry spiny scrub forest of southwestern coastal Madagascar and contains three geographic races, that is, subspecies. The carapace is high domed, and the plastron has a hinge between the humeral and pectoral scutes. Although both features may protect against predation, the ability to close the shell also serves to reduce water loss as these tortoises estivate semi-buried in leaf litter and soil for nearly nine months. They emerge with the first rains to feed on young leaves and to reproduce. Females lay a single egg each year. This low reproductive rate is a major factor in their endangered status.

Flat-tailed tortoise (*Pyxis planicauda*)

♀♀ 11–14 cm; ♂♂ 9–15 cm adult SCL

This species is another dwarf tortoise; it has a domed, rectangular carapace and ranks as the second most endangered tortoise in the world, confined to an area of about 5,000 km^2 of deciduous forest on the west-central coast of Madagascar. Within that area, they inhabit patches of forest on sandy soil. The flat-tailed tortoise appears to spend between six and seven months (approximately May through October) estivating. They stay partially buried and covered with leaf litter to survive the extreme aridity of the habitat. Only with rains and cooling temperatures do they emerge and begin feeding on fruit and younger plant growth.

Tropical Asian Tortoises

The following tortoises have a patchy distribution (Fig. 7.5) from extreme western India eastward to extreme southwest China and southward through Malaya into Sundaland (Borneo, Sumatra, and other smaller islands. Although some species are adapted to arid conditions, a few live in wet tropical forests, a lifestyle shared only with a single African species and two South American species.

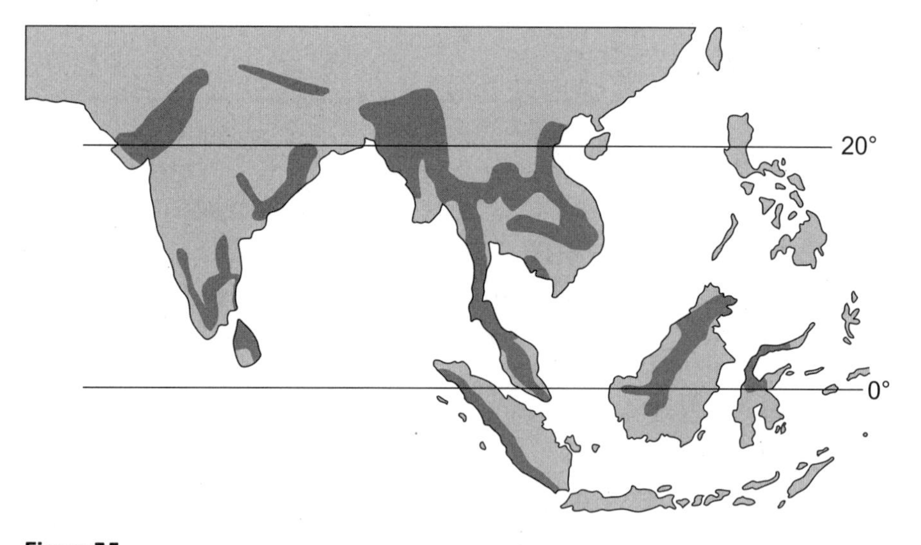

Figure 7.5
Generalized distribution of the extant tortoise species of subtropical and tropical Asia.
Data from Turtle Taxonomy Working Group (2021)

Indian Star Tortoise *(Geochelone elegans)*

♀♀ 20–29 cm, exceptional to 41 cm; ♂♂ 22–38 cm adult SCL

This moderate-sized species (Plate 1) lives in three widely separated dry forests in peninsular India, Thar and the Eastern Ghats, and in northern Sri Lanka. It has a domed carapace commonly with low pyramidal dorsal scutes, each highlighted by a thin yellow starburst pattern (that inspired its common name). They occupy a range of habitats from natural grasslands to human-made plantations and hedgerows, and from sea level to as high as 900 m.

Burmese Star Tortoise *(Geochelone platynota)*

♀♀ 20–46 cm & ♂♂ 17–30 cm adult SCL

This moderate-sized species (Fig. 5.2) occurs in the dry forest and grasslands of Myanmar's Central Dry Zone. Like its close relative the Indian Star Tortoise, its pyramidal scutes have a yellow starburst pattern, for which it is named. Hatchlings have markedly domed carapaces that slowly elongate with growth. This species was functionally extinct in the early 21st century, but captive breeding and reintroductions of thousands of subadults have restored it in two protected wildlife sanctuaries areas within its original distribution.

Yellow-headed or Elongated Tortoise *(Indotestudo elongata)*

♀♀ 18–32 cm; ♂♂ 24–39 cm adult SCL

This moderate-sized species (Figs. 2.1*B*, 2.6*F*) lives in dry to moist deciduous forests from Nepal through Myanmar into Thailand and the northern Malayan Peninsula. Named for its elongate and rectangular and usually slightly convex shell typical of the genus, this species prefers dense vegetation and may take refuge in streamside forests during the dry season when forage is scarce. Their home ranges are relatively large. Its disconnected populations are restricted by geology and human development to areas such as a single national park in Nepal.

Forsten's or Sulawesi Tortoise *(Indotestudo forstenii)*

♀♀ 11–25 cm; ♂♂ 12–27 cm adult SCL

This moderate-sized species (Fig. 9.3) lives in the Acacia scrub forests of northern Sulawesi, Indonesia. It also tolerates dry, disturbed habitats and occurs in secondary growth forests, including palm, rubber, and teak plantations as well as alongside village streams. Females lay only one or two eggs in a clutch but continue to lay throughout the year. These tortoises are critically endangered.

Travancore Tortoise (*Indotestudo travancorica*)

♀♀ 19–30 cm; ♂♂ 21–33 cm adult SCL

This moderate-sized species (Plate 21) lives in open areas of dry deciduous and evergreen forests of the Western Ghats, India. They spend most of their time on the edges of evergreen forests in bamboo-lantana grasslands. Edge habitat may be optimal for maintaining body temperature as they shuttle between warmer grassland and cooler woodland. In the midday heat, Travancore Tortoises are found taking shelter in tree buttresses, logs, and pangolin burrows.

Asian Giant Tortoise (*Manouria emys*)

♀♀ 50–58 cm; ♂♂ 46–60 cm adult SCL

This large species (Plate 3) occurs in or adjacent to water in tropical evergreen forest and bamboo forest of western Myanmar, Malayan Peninsula. It is also found in western Sumatra and central Borneo. This tortoise species and the other one in its genus—*M. impressa*—guard their nests against predators, a behavior known in only this genus and in *Gopherus* (Chap. 4). The female positions her body over the nest to keep predators from digging into it and shoves predators away.

Impressed Tortoise (*Manouria impressa*)

♀♀ unknown to 36 cm; ♂♂ unknown to 29 cm adult SCL

This moderate-sized species (Fig. 6.3) lives in evergreen and deciduous forests of Myanmar and northern Southeast Asia from about 600 to 1,300 m above sea level. Both species of *Manouria* (this one and *M. emys*) have heavy-boned, elongate rectangular shells with flat tops, although the Impressed Tortoise is more flattened. Both are also decidedly forest inhabitants, preferring "cool" and moist, litter-covered forest floors. Aside from fallen fruit and some invertebrates, mushrooms are their major food source, especially during the rainy season when mating peaks.

North American Tortoises

All North American tortoises are in the genus *Gopherus*. The members of this genus and of *Manouria* each represent the most primitive and oldest lineages. Their ancestry extends back to the Early Eocene. They diverged only slightly from their original tortoise ancestors. Since they first appeared, Gopher Tortoises have remained North American tortoises and have been found on the continent's southern edge from coast to coast (Fig. 7.6). Most species, both extinct and living, have muscular forearms and hands with broad, heavy claws for digging.

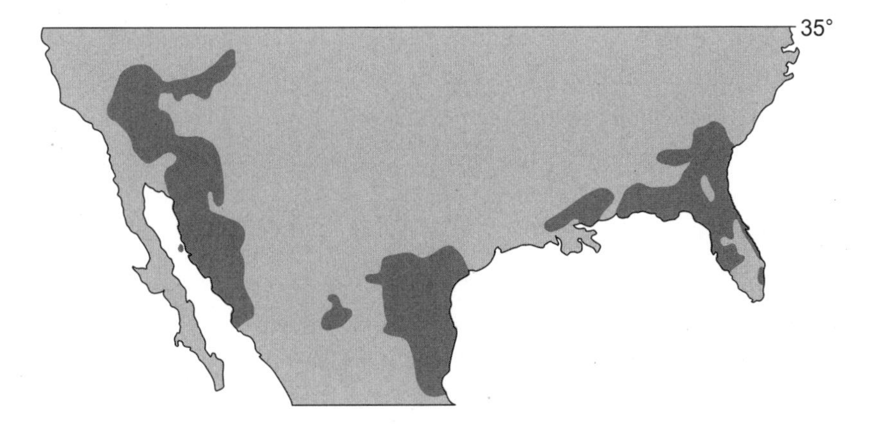

Figure 7.6
Generalized distribution of the extant species of North American tortoises. All six species are members of the genus *Gopherus*. Data from Turtle Taxonomy Working Group (2021)

Mojave Desert Tortoise (*Gopherus agassizii*)

♀♀ 12–29 cm; ♂♂ 18–37 cm adult SCL

This moderate-sized species (Figs. 2.6*B*, 4.3*A-D*, 9.4, 10.4; Plates 6, 11) with a flattened, rectangular shell is native to the Mojave Desert of California, Nevada, Utah, northwestern and southwestern corners of Arizona, and a bit of the Sonoran Desert in Southern California The latter areas have a variety of desert habitats and rainfall schedules. The tortoises live only in areas with shrubs or small trees, where they excavate pallets and burrows in the shade of the vegetation or rock outcrops. Human activities have severely altered or destroyed prime tortoise habitats. Most, if not all, Mojave Desert Tortoises populations now have densities below replacement rates.

Berlandier's Tortoise (*Gopherus berlandieri*)

♀♀ 13–21 cm; ♂♂ 13–24 cm adult SCL

This small species (Figs. 4.7, 6.4) with a domed rectangular shell is a resident of the semidesert thornscrub landscape of the southern half of Texas and adjacent northeastern Mexico. The soils range from sand to clay to gravelly mineral deposits (caliche). It does not dig long burrows, although it regularly excavates shallow depressions ("pallets"; Chap. 6) beneath shrubs to avoid the heat of the sun and reduce the chill of winter cold spells.

Sinaloan Thornscrub Tortoise (*Gopherus evgoodei*)

♀♀ and ♂♂ to 26 cm adult SCL

This moderate-sized species with a domed, rectangular shell has a restricted range in tropical deciduous forests and thornscrub of Sinaloa, Mexico. Sinaloan Thornscrub

Tortoises take shelter in packrat middens, dry cacti, and animal burrows. Until recently (2016), this population was identified as *G. morafkai*. It occurs together with this Sonoran *Gopherus* lineage, *G. morafkai,* in a narrow area where they interbreed, yet each maintains a unique genetic identity. *G. evgoodei* can be distinguished by its brown carapace with a distinctly orange hue.

Bolson Tortoise (*Gopherus flavomarginatus*)

♀♀ 31–39 cm; ♂♂ 26–35 cm adult SCL

This moderate-sized species with a flat, rectangular shell profile is the largest species of North American tortoise. It now inhabits the grassy margins of floodplain basins in the arid grasslands of the Bolsón de Mapimí in northeastern Mexico, although fossils indicated that it once occurred as far north as the Big Bend in Texas. Like its closest relative, *G. polyphemus*, it digs extensive burrows several meters long, spending only about 1% of its time on the surface foraging on grasses and herbs.

Sonoran Desert Tortoise (*Gopherus morafkai*)

♀♀ 18–30 cm; ♂♂ 18–30 cm adult SCL, ♂♂ rarely to 38 cm

This moderate-sized species inhabits the Sonoran Desert, south and east of the Colorado River, in the southwestern United States and adjacent Sonora, Mexico (Figs. 4.4, 6.7; Plate 5). *G. morafkai* shares most aspects of its ecology as a burrowing tortoise with *G. agassizii*, its close relative on the other side of the Colorado River, although the *G. morafkai* prefers slopes and rocky hillsides. Until recently (2011), it was identified as *G. agassizii*, but genetic analyses showed it to be a separate Sonoran lineage. *G. morafkai* has a narrower shell and more pear-shaped carapace than *G. agassizii.*

Gopher Tortoise (*Gopherus polyphemus*)

♀♀ 23–31 cm, maximum reported 39 cm; ♂♂ 18–27 cm adult SCL

This moderate-sized species (Figs. 6.2, 6.6; Plate 22) occupies sandhill pine-oak savannas, from central peninsular Florida to southern Georgia and the Gulf Coast to eastern Louisiana. *Gopherus* species, aside from the Galápagos giants and European *Testudo*, are likely the most studied tortoise species, in a large part owing to development pressures for its upland habitat by the ever-expanding human population of Florida and the southwestern United States.

South American Tortoises

All South American tortoises (Fig. 7.7) are members of the genus *Chelonoidis*. Their closest relatives appear to be the African tortoises of the *Centrochelys sulcate* group.

Their African ancestors are another example of ability of large tortoises to disperse by floating across marine barriers, in this instance, the Atlantic Ocean (Chap. 8). Admittedly, in the Oligocene, the Atlantic Ocean between Africa and South America was narrower than today, although still hundreds of miles wide. Their initial American home appears to have been northern South America, from which they spread southward through much of continental South America east of Andes and into the islands of the West Indies and the Galápagos.

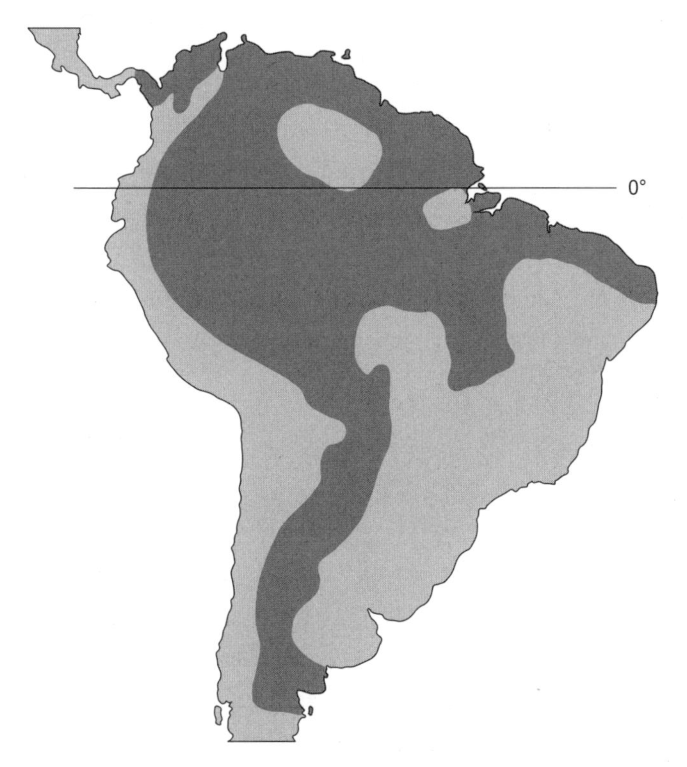

Figure 7.7
Generalized distribution of the extant species of South American tortoises. All four species are members of the genus *Chelonoidis*. Data from Turtle Taxonomy Working Group (2021)

Red-footed Tortoise (*Chelonoidis carbonarius*)

♀♀ 21–44 cm; ♂♂ 18–43 cm adult SCL

This moderate-sized species (Fig. 4.2) with a rectangular shell inhabits the savannas and their forest edges of northern South America. It includes an isolated population in southwest Brazil that lives in forested coastal shrub habitat. This tortoise is commonly raised in captivity and kept as a pet; some captive individuals have attained 50–51 cm SCL. Because of its captive popularity, it has been introduced in multiple places, including Caribbean islands such as St. Croix.

Chaco Tortoise (*Chelonoidis chilensis*)

♀♀ 18–28 cm; ♂♂ 15–22 cm adult SCL

This moderate-sized species (Plate 2) has a rectangular shell. Its wide distribution in South America includes the mixed savanna of the pampas of western Paraguay, Argentina, and Bolivia, and extends into the dry forests (Dry Chaco) and dry thornscrub (Monte) ecoregions. Chaco Tortoises have been observed using the burrows of the large rodent, the Patagonian mara (*D. patagonium*).

Yellow-footed Tortoise (*Chelonoidis denticulatus*)

♀♀ 29–71 cm; ♂♂ 35–82 cm adult SCL

This moderate-sized to giant species (Plate 25) lives in a variety of forest types and is often found in adjacent savannas of the northern half of South America (mainly the Amazon Basin. Unlike its close relative, the Red-footed Tortoise, *C. denticulatus* is associated with tropical and subtropical wet forests. *G. denticulatus* is estimated to have diverged from a common relative it shares with *G. carbonarius* from 2.2–4.0 mya, when rainforest fragmentation confined the latter to its preferred savannah habitats.

Tortoises of the Galápagos Islands

Galápagos tortoises have a variety of shell shapes, ranging from nearly hemispherical domed carapaces through domed rectangular to saddle-backed ones that have anteriorly upward curved carapaces (Fig. 8.3). All Galápagos tortoises are also commonly labeled as giants, even when some individuals never attain carapace lengths greater than 60 cm; nevertheless, the maximum length of a species determines their entrée into "giant" status. Until recently, each genetically distinct population was recognized as a species. Now they've been assigned subspecific status thanks to a more detailed genetic study coupled with fossils from the Caribbean (Chap. 8). The subspecies are genetically distinct with one per island, except for Isabela and Santa Cruz, and possibly San Cristóbal, where multiple subspecies occur, each associated with a particular volcano. Three subspecies have gone extinct due to human effects on their populations over the past century (Figs. 9.1, 9.2). There is continued sporadic poaching of Galápagos Tortoises as well as loss of nests to fire ant, rat, and feral pig predation, prompting the need for continued management (Figs. 5.1, 10.2).

Pinta Giant Tortoise (*Chelonoidis niger abingdonii*)

Extinct

♀♀ unknown; ♂♂ 44–92 cm adult SCL

This large species with a saddleback carapace formerly lived in the rocky cactus scrub of Isla Pinta. The last known individual was a male called "Lonesome George" who died in 2012, marking the extinction of the subspecies. While George was part of a captive breeding program on the island of Santa Cruz, he failed to reproduce and left no known descendants. Recently, some individual tortoises with Pinta ancestry have been discovered on Isla Isabela near Volcán Wolf, the native habitat for the Volcán Wolf Giant Tortoise (*C. n. becki*).

Volcán Wolf Giant Tortoise (*Chelonoidis niger becki*)

♀♀ 55–87 cm; ♂♂ 46–105 cm adult SCL

This large, dome-shelled species has a rectangular domed carapace that approaches a saddleback shape. It's the only species of giant tortoise known to occur naturally in the thorny scrub and high-elevation open woodland on the slope of Volcán Wolf, Isla Isabela, where it feeds on stiff grasses and shrub leaves. The total area it occupies is about 246 km^2. Despite more than half its population disappearing over the past two centuries, it's still common in some areas around Volcán Wolf. It shares the island with four other subspecies of *Chelonoidis niger*. Each has a distinct distribution centered on one of the five still-active volcanoes.

San Cristóbal Giant Tortoise (*Chelonoidis niger chathamensis*)

♀♀ 60–90 cm; ♂♂ 75–110 cm adult SCL

This large species with a domed carapace occupies the grassland and scrub forest of Isla San Cristóbal, the oldest island in the Galápagos. Females are more domed, while males are flatter in profile, sometimes with a slight saddle shape. It's the only species of tortoise known to occur on San Cristóbal. Historically, it occurred throughout the island, but today it is limited to a 52-km^2 area in the northeast part of the island. Juveniles today live in the warmer, lowland coast and migrate upland as adults at 10–15 years of age. A recent study suggests that the Isla San Cristóbal tortoises living today are distinct from the described *C. n. chathamensis*, which if confirmed, will warrant a new name for the living subspecies.

Santiago Giant Tortoise (*Chelonoidis niger darwini*)

♀♀ 55–95 cm; ♂♂ 75–140 cm adult SCL

This large species has a domed carapace with a big anterior opening. It regularly migrates between the scrub and grassland and higher-elevation open woodlands of Isla Santiago. Charles Darwin collected a juvenile Santiago Giant Tortoise during his visit to the Galápagos on the *Beagle* voyage and took it home to England. This tortoise pet disappeared from records, but it was rediscovered in 2010 in a jar of alcohol in London's Natural History Museum, labeled on its plastron as Darwin's pet. Human harvest and collection of these tortoises reduced their population from an estimated 24,000 adults to approximately 600 by the 1970s. This population is today augmented through captive breeding and reintroductions (Chap. 10).

Cerro Fatal Giant Tortoise (*Chelonoidis niger donfaustoi*)

♀♀ to 114 cm; ♂♂ to 141 cm adult SCL

This large species with a domed to rectangular carapace inhabits the scrub and grassland of eastern Isla Santa Cruz. This unique population of tortoises was only recently recognized as a distinct subspecies. It was named in honor of Fausto Llerena Sánchez, who worked for 43 years on giant tortoise conservation as a park ranger in the Galápagos National Parks.

Pinzón Giant Tortoise (*Chelonoidis niger duncanensis*)

♀♀ to 80 cm; ♂♂ to 94 cm adult SCL

This large, dome-shelled species was extirpated from its original home of Isla Pinzón but was introduced on the small, rat-free Isla Santa Cruz from the captive-rearing colony at the Galápagos Research Center. Isla Pinzón tortoises were reduced to just 100–200 adults by 1959, mostly due to predation by introduced black rats. The subsequent captive rearing and repatriation program, coupled with the eradication of black rats, has resulted in a larger, reproductively active population of Pinzon Giant Tortoises.

Sierra Negra Giant Tortoise (*Chelonoidis niger guntheri*)

♀♀ 60–74 cm; ♂♂ 62–100 cm adult SCL

This large species is now largely confined to the scrubby woodland on the slope of Volcán Sierra Negra, Isla Isabela, owing to human agriculture in the higher elevations. It has two shell variations; domed individuals tend to graze on herbs, while flattened individuals ("aplastados") favor shrubs. Its range used to connect with the range of *C. n. vicina*, which is also its closest relative, but their ranges are now disjunct after intensive human harvest.

Española Giant Tortoise (*Chelonoidis niger hoodensis*)

♀♀ to 76 cm; ♂♂ to 91 cm adult SCL

This moderately large saddleback species with yellow jaw and throat markings inhabits the desert scrub forest of Isla Española (Fig. 2.6*D*). On the island, tortoises are concentrated in the central zone east to Gardner Bay, occupying about 13% of the habitat considered suitable. The population of Española Giant Tortoises collapsed from human exploitation to a mere 14 individuals by the 1960s. They were rescued thanks to a breeding program that successfully repopulated the island with some 800 tortoises, although their genetic diversity is low.

Volcán Darwin Giant Tortoise (*Chelonoidis niger microphyes*)

♀♀ 55–95 cm; ♂♂ 74–103 cm adult SCL

This large, dome-shelled species moves seasonally between lower scrublands to higher-elevation woodland slopes of Volcán Darwin, Isla Isabela. It shares the island with four other subspecies of *C. niger*, each with a distinct range on one of the five still-active volcanoes. The Volcán Darwin Giant Tortoise ranges from the western flank of the volcano, the more vegetated side, down to the coast. About 800 individual tortoises remain from an estimated population of 14,500 before human harvest devastated their original population.

Floreana Giant Tortoise (*Chelonoidis niger niger*)

Extinct

♀♀ ??–?? cm; ♂♂ ??–96 cm adult SCL

This large tortoise formerly lived on Isla Floreana. Until recently, this tortoise was known as *G. elephantopus* and likely should retain that specific epithet. From an estimated 8,000 Floreana Giant Tortoises in the early 1800s, the Floreana population collapsed by 1850 because of human harvest. But some individual tortoises with Floreana ancestry have been discovered on Isla Isabela in the vicinity of Volcán Wolf, the native habitat for the Volcán Wolf Giant Tortoise (*C. n. becki*). A few captive adults show some Floreana ancestry. *C. n. niger* genes in other populations are likely the result of early 19th-century seamen moving them from island to island.

Fernandina Giant Tortoise (*Chelonoidis niger phantasticus*)

Near Extinct

♀♀ ??–45 cm; ♂♂ ??–88 cm adult SCL

This large saddleback tortoise is native to the scrub forest on the slope of Cerro Pajas, Isla Floreana. Its scientific name originated from the male's dramatic saddleback shape and the extreme curvature of its marginal (carapace edge) scutes. The lower jaw and throat of this subspecies have yellow markings. The Fernandina Giant Tortoise was assumed to be extinct, as it was known from only a single 1906 specimen; however, a living adult domed female tortoise was found on the lower edge of Volcán Cerro Pajas in 2019. Nuclear DNA analysis shows her and the 1906 individual to be from the same lineage and distinct from all others, whereas other data (mitochondrial DNA) shows genetic divergence. Whether *C. n. phantasticus* is extinct or living awaits further study.

Western Santa Cruz Giant Tortoise (*Chelonoidis niger porteri*)

♀♀ 65–118 cm; ♂♂ 85–151 cm adult SCL

This large tortoise with a domed to elongate-domed carapace occupies the grassy woodlands of Isla Santa Cruz, as well as the arid zone containing cacti such as prickly pear and the tall *Jasminocerus* cactus (Fig. 6.5, Plate 4). Their eggs experience a high (more than 50%) rate of mortality from feral pig predation, trampling by donkeys, and attacks by fire ants on hatchlings (Chap. 6). DNA analysis shows this subspecies to be closely related to the Fernandina Giant Tortoise (known from just one deceased and one living individual).

Volcán Alcedo Giant Tortoise (*Chelonoidis niger vandenburghi*)

♀♀ 65–107 cm; ♂♂ 85–148 cm adult SCL

This large tortoise (Fig. 2.6*A*, Plate 12) with a dome-shaped carapace lives on dry forest slopes of Volcán Cerro Alcedo, Isla Isabela, and migrates seasonally upward to wetter woodlands. It shares the island with four other subspecies of *C. niger*, but with distinct ranges, each centered on one of the five still-active volcanoes. The Volcán Alcedo Giant Tortoise has one of the largest remaining populations of Galápagos tortoises, last estimated at more than 6,000 individuals. That's still small compared to an estimated 38,000 individuals before humans settled the Galápagos Islands three tortoise generations ago.

Cerro Azúl Giant Tortoise (*Chelonoidis niger vicina*)

♀♀ 54–80 cm; ♂♂ 62–125 cm adult SCL

This large tortoise with a dome-shaped carapace lives in thorny scrub on the slopes of Volcán Cerro Azúl, Isla Isabela. Some older males develop a saddleback carapace. It shares the island with four other subspecies of *C. niger*, but with distinct ranges, each centered on one of the five still-active volcanoes. Its range used to connect with the range of *C. n. guntheri*, which is also its closest relative. Their ranges are now separated owing to past human harvest.

Santa Fe Tortoise

Extinct, not described.

CHAPTER 8

Ancestry and Evolution

Tortoise Origins

Turtle Origins

During the Late Permian 263–260 mya (Fig. 8.1), in what would later be called the Karoo Basin of South Africa, a strange-looking lizardlike *Eunotosaurus africanus* foraged along swampy shorelines in search of its prey. About 30 cm long from nose to tail tip, this reptile had a small head on a modest-length neck of six vertebrae protruding from a broad, flattened body with stout limbs and a long tail. It was discovered in 1892, and in 1914, an eminent anatomist proposed it as the most recent common ancestor of living turtles.

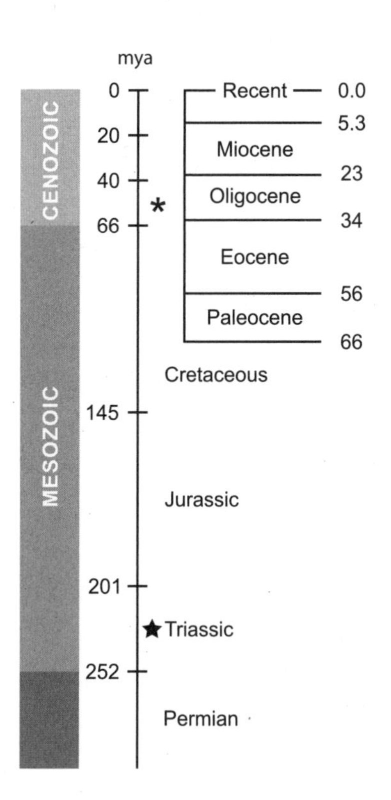

Figure 8.1
Geologic time chart illustrating the vastness of time occupied by turtles from their earliest ancestors (star) appearing in the Middle to Late Triassic ~252 mya to the origin of tortoises (asterisk) during the Paleocene-Eocene transition to the present. The Pliocene (5.3–2.58 mya), Pleistocene (2.58 mya to 12 kya), and the Holocene (12 kya to present) are too brief to be accurately portrayed. The timescale is equal throughout the graph. Data derived from the Geological Society of America's Geological Time Scale, version 5

Although *E. africanus* did not have a shell, its widened, horizontal ribs and few (nine) trunk vertebrae suggested a close relationship to modern turtles. It had already lost the ability to expand and contract its ribs, thus shifting to a turtle-style respiratory mode. Nevertheless, its status as a turtle ancestor was declared invalid in 1956 by another eminent anatomist-paleontologist, and so it remained until earlier this century. The continual discovery of fossils of similar anatomical reptiles and more early turtles with complete carapaces eventually led to the resurrection of *Eunotosaurus* as the earliest known ancestor of turtles. Still, analyses continue to cast doubt on its position in the turtle lineage, a recent study noting a cranial (skull) structure distinct from what you'd expect in a turtle ancestor.

Acquiring Their Shells

Regardless of whether *E. africanus* is the right culprit for the "stem" turtle (ancestor that gave rise all turtle descendants, classified in Pantestudines) or whether the honor should be bestowed on some other yet-to-be-discovered species, turtles are an incredibly old group of vertebrates. The incomplete shells of these Late Permian and Early Triassic turtlelike reptile candidates were way stations along the path of turtle evolution. Acquiring a full shell would require overhauling breathing, moving, and feeding mechanisms. Plus, turtles were to dip their proverbial toes in the water—as semiaquatic and aquatic species—before tortoises came to be.

As the ribs broadened in stem turtles, they started to constrain the rib movements needed for breathing (Chap. 3). At the same time, some muscles (the hypaxial muscles) changed to stop supporting the turtle's trunk and instead became solely dedicated to ventilating their lungs. Fossil evidence shows that these necessary changes to accommodate a complete shell were already taking place in reptiles like *E. africanus*. Other Triassic turtlelike reptiles—such as Germany's *Pappochelys,* China's *Odontochelys* and *Eorhynchochelys,* and Argentina's *Palaeochersis* and *Waluchelys*—had various subsets of turtle features, like a beak and a plastron (bottom shell), which showed up complete before the top shell. Their discovery locales and bone structure indicated a semiaquatic (amphibious) lifestyle on the margins of lakes, wetlands, or ocean coastlines.

A Late Triassic fossil bed in Germany (dating to 221.5–205.6 mya) yielded remains of the turtle *Proganochelys quenstedti* (Fig. 8.2), thought to be in a sister group to all turtles (closest relatives through a common ancestor). The oldest known stem turtle with a complete shell, *Proganochelys* has been variously described as terrestrial and aquatic. Its broad, somewhat flattened shell (carapace length to ~1 m), wide head, and long tail imply the general appearance of today's Snapping Turtles (*Chely-*

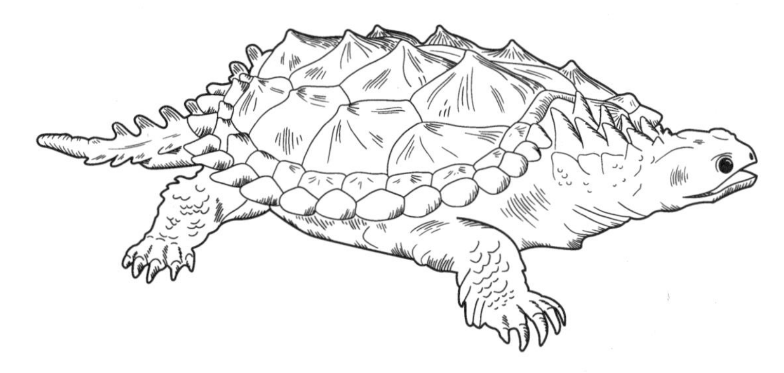

Figure 8.2
Proganochelys quenstedti, the second-oldest known species of the true turtles (Testudines). It is represented by multiple fossils from the Late Triassic of Germany, Greenland, Switzerland, and Thailand. Carapace length was 60–70 cm. Body and limb anatomy suggests that it was semiterrestrial, like snapping turtles today. Sketch of the plaster model Gaffney (1990, Fig. 4)

dra serpentina) that are mostly aquatic but regularly embark on overland journeys. It may have also been the first turtle able to retract its neck into its shell by pulling it in laterally (sideways), thus giving the shell a protective role. The beaked *Proganochelys* had teeth, although not on its jaws; the cone-shaped teeth lined up in several rows on its palate (roof of its mouth). As teeth gave way to sharp, keratinous beaks in turtles, the palate teeth were the last to go.

True Turtles

The oldest known animal considered a "true" turtle with a complete shell is *Proterochersis robusta* from the same Late Triassic fossil beds in Germany as *Proganochelys*. It had the major characteristics by which zoologists define turtles, including a fully developed and joined carapace and plastron, plus a high-domed carapace. The domed shell suggests a terrestrial or possibly semiaquatic lifestyle, although shell shape is not always a reliable predictor of turtle lifestyle, as there are a few domed aquatic turtles such as Mud and Musk Turtles (family Kinosternidae) and at least one startlingly flat tortoise (Pancake Tortoise).

During the Late Jurassic and into the Cretaceous, fossil turtles evolved several characteristics seen in modern turtles. These species are grouped and labeled as Mesochelydia, denoting their "middle position" in the evolution of modern characteristics; their ancestry witnessed the development and refinement of a few major biological features of turtles. Among the variety of fossil genera of turtles, three lineages arose that survived to modern times. One of those—Meiolaniformes—contained several species of terrestrial, horned *Meiolania* of the southwest Pacific and Australia. These mostly large turtle species survived nearly into the present, with the last species disappearing when humans arrived in New Caledonia (Chap. 9).

The other two lineages, Pleurodira (side-necked) and Cryptodira (S-necked), had aquatic representatives (Appendix Table A.1) and have survived to the present with numerous descendants across the world (Chap. 7). The side-necked pleurodires are currently confined to the Southern Hemisphere (Africa, Australia, and South America). Cryptodira occur worldwide, including all the major oceans except the Antarctic. Most living turtles including tortoises (family Testudinidae) are cryptodires, which fold their necks straight back into their shells. If you look at a cryptodire tucked into its shell, you'll see its nostrils peeking out between the front legs (see Fig. 2.2*A*).

Cretaceous Changes

Fast-forward into the Late Cretaceous. All known cryptodire species were aquatic or semiaquatic; terrestriality had waned. The worldwide climate was deteriorating from centuries of volcanic eruptions in the Deccan area of what is now India as it passed over a hot mantle plume (the Réunion hotspot) during India's movement northward to collide with Asia. Ongoing climate change was capped by the massive Chicxulub meteor strike 66 mya in an area that is now the Yucatán Peninsula of Mexico. The resulting ecocatastrophe caused the loss of more than 60% of the world's biodiversity, including the large dinosaurs. Yet all three turtle lineages survived this mass extinction event, a testimony to the durability of turtle anatomy and physiology.

It is possible that adapting to the deteriorating climate was a factor contributing to the return to a terrestrial lifestyle of the tortoises. Still, within the Cryptodira lineage, only the tortoise family (Testudinidae) became fully terrestrial. In a case of convergent evolution, a single genus (*Terrapene*, the box turtles) in the family Emydidae was to later become totally terrestrial. So, the terrestrial adaptations of tortoises are a unique and central part of their identity. That said, all turtle species past and present have a terrestrial aspect of their lifestyle because they must nest on land. Use of terrestrial habitats for nesting has been a part of their reproduction since at least the earliest known true turtle, *Proterochersis*.

Tortoise Origins

Interpreting the identity and relationships of fossilized organisms is challenging because many animals do not leave a fossil record, or at least not one that gets discovered. When fossils are found, they're rarely the whole skeleton. The accuracy of fossil interpretations depends on variables like location, context, and prevalence of fossils of that species, making for spurious gaps in the fossil record. But biologists studying tortoises are relatively lucky. The turtle fossil record is exceptional compared to most other vertebrate groups. Shells or pieces of shell turn up in the fossil record, even occasionally limb bones and skulls. The fossil record illuminates the origins and extinctions of the different tortoise lineages over time.

Still, constructing evolutionary trees (phylogenies) from the fossil record is fraught with difficulty. A fossil marks just a snapshot in time of the death and subsequent preservation of a single organism. As such, it is a point on the branching continuum of evolution over millions of years, for our purposes a continuum of turtles to tortoises. Molecular data are particularly useful in discerning the relationships among species themselves, whereas fossils best illuminate minimum points at which divergence occurred, that is, splitting of a group into two descendent groups. As you go deeper into the fossil record toward points of origin, however, fossils become increasingly rare, so divergence dates must be viewed as minimum estimates. Fossils of earlier members of an ancestral species may not be easily unearthed, or environmental conditions may not have supported fossilization.

Until the advent of molecular techniques, tortoise evolution was interpreted solely from fossils, looking through a geographic lens. Tortoise fossils on each continent were believed to be the direct ancestors of the modern tortoises we see on Earth today. For terrestrial organisms like tortoises, it made sense to assume that evolution happened independently over long periods on continents that were separated by oceans. The resulting picture was one of lineages pertaining to each continent. Now, the picture of the tortoise tree of life is changing as DNA evidence brings unexpected relationships to light. The molecular evidence indicates a lot more migration and dispersal, even across oceans, leading to a more complex picture of tortoise origins. Here, we aim to report the most recent and strongest hypotheses of tortoise evolution based on the synthesis of fossils and molecular studies, knowing that many uncertainties may be resolved over time through ongoing research.

Stem Tortoises and Lineages

Relative to other families of turtles, tortoises are a young group. They arose from a lineage of geoemyid turtles (or "Old World" Pond Turtles), a large group of turtles that today includes 70 species living mostly in Asia ranging from semiaquatic to fully aquatic. Despite the earliest known and most primitive tortoise fossils to date coming from North America and France (see below), molecular studies situate the origins of tortoises from geoemydid in Asia. Fossils representing the first Asian tortoises have yet to be uncovered. They would have been closely related to aquatic turtles identified from fossils found in China: Early Paleocene *Elkemys* and later *Hokouchelys* (66 mya and upward).

Onto Land

The origin or transition from the lifestyle of aquatic geoemydid turtles to terrestrial tortoises was underway by the Late Paleocene (~58 mya) at the beginning of the exceptionally warm Paleocene-Eocene Thermal Maximum (PETM) period. A basal (primitive-looking) Paleocene turtle from Mongolia, for example, looked to be semiterrestrial, living on land but likely feeding underwater. During the PETM, the average global temperature rose 5°C–8°C above today's global temperatures. Even though this may sound like a negligible change, it resulted in the melting of all polar ice caps. The new warm, wet conditions may have facilitated the evolution of the earliest tortoises (stem tortoises) in the Northern Hemisphere.

Across the Globe

During the Eocene (starting 56 mya), the main lineages of tortoise emerged that would ultimately spawn all the tortoises we see today. Those included (1) the oldest lineage of Asian *Manouria* (with only two living species today); (2) the almost equally old lineage of Gopherini (including today's *Gopherus*) in North America; and (3) the Testudininae (which left a global fossil record spanning multiple continents) that by the Late Eocene had split into the Testudona lineage and the Geochelona lineage. Ultimately, the Geochelona lineage would leave its mark as the lineage that includes nearly all tortoises remaining on Earth today.

Modern and fossil representatives of Testudona and Geochelona are peppered across continents, suggesting a complex suite of tortoise migrations from the Eocene onward, which may have encouraged diversity. During the Late Eocene, tortoises included at least 23 species. Diversity decreased during the Oligocene into Miocene and then picked up again in the Middle Miocene. Plate tectonics set the stage for global movements of tortoises—during periods with land connections between Africa, Asia, and Europe—although they apparently moved not only by land but also by sea. Most living and recently extinct tortoises have roots (as members of Geochelona) in Africa, where several waves of dispersal took place, including to South America.

Ocean Dispersals

Colonizing Continents

Tortoises dispersed throughout the Northern Hemisphere during the Early Eocene and had become widespread by the Middle Eocene. Dispersing tortoises may have taken advantage of some of the land corridors available at the time, but they are also known to weather long-distance ocean floats, thanks to buoyant bodies and a physiology geared to resilience. To those unfamiliar with tortoise hardiness (Chap. 3), ocean floats sound improbable, but the capacity of tortoises to survive ocean crossings is well documented.

A female Aldabra Giant Tortoise (*Aldabrachelys gigantea*; Plate 20) walked ashore onto a beach in Dar es Salaam, Tanzania in December of 2004. She likely came from Aldabra Atoll, approximately 740 km away. The age of the dozens of barnacles that had colonized her legs suggested that she had been at sea for at least several weeks, apparently having ridden the South Equatorial Current from Aldabra to Tanzania. Although this is the first direct evidence of a tortoise surviving an ocean crossing, the presence of tortoises on distant oceanic islands demonstrates that tortoises disperse to islands by floating or rafting on debris.

History also records comments on floating tortoises. The journal of David G. Farragut, a young officer aboard the 1813 USS *Essex,* wrote that "they had thrown overboard several hundred Galápagos terrapins. The appearance of these turtles in the water was very singular: they floated as light as corks, stretching their long necks as high as possible." In 1923, naturalist William Beebe tested the buoyancy of tortoises by throwing one over the side of a ship and watching it bob in the water. Other naturalists since have observed tortoise floating, such as Gopher Tortoises floating high in the water with heads out and swimming short distances by paddling their legs. And floating across oceans is not just a recent phenomenon among tortoises. It explains how they reached islands long before humans came on the scene (see the Continent by Continent section).

Body Size Evolution

During the evolution of tortoises, some became gigantic. One hypothesis for animal species evolving toward large body sizes is the phenomenon of "island gigantism," where species that floated to predator-free islands tended to evolve independently to grow much bigger than their mainland cousins. Examples that have been put forth to substantiate the theory are the Giant Hissing Cockroach of Madagascar, the Giant Moas of New Zealand (extinct), and giant tortoises. However, gigantism in tortoises cannot be explained by the island effect.

Fossils and genetic data indicate that gigantism in tortoises evolved repeatedly, including in at least two species that lived in what is now the southern United States through the Pleistocene—*Hesperotestudo crassiscutata* and *Gopherus hexagonatus*—each longer than a meter. Giant tortoises evolved multiple times on the mainland of Africa, the Americas, Asia, and Europe, suggesting that gigantism was happening before tortoises reached islands. Indeed, larger tortoises may have been better equipped to colonize the islands because of higher tolerance to fasting and dehydration. At least for some instances of gigantism, climate change may have played a role, for example, in the evolution of *Titanochelon* spp. during the Late Miocene as climate got drier and cooler, with larger tortoises benefiting from physiological advantages of lower ratios of body surface area to volume.

The oldest known instance of gigantism (the tortoise *Gigantochersina ammon*) took place from the Early Oligocene through Pliocene of Africa, coinciding with the increasing diversity in tortoises. Not long after, gigantism appeared in Oligocene tortoises in Europe (*Taraschelon gigas*). Additional giant tortoises cropped up in the Miocene, Pliocene, and Pleistocene, ultimately occupying all continents where tortoises live. In the Geochelona lineage, tortoise body size spanned an impressive range from small species to the biggest tortoise that ever lived on Earth, *Megalochelys sivalensis* from Asia, with a carapace length up to 2.7 m (the size of a smart car).

Today, giant tortoises remain in the wild only in two locations, on the Galápagos Islands and Indian Ocean islands, having mostly gone extinct during the Pleistocene when humans came into their habitats. Indeed, an analysis of tortoise body size distribution over the past 23 million years shows a dramatic reduction in body size, first on mainland landmasses (from ~2.588 to 0.781 mya) and then on islands (from ~0.126 to 0 mya), in keeping with human migration patterns.

Also, some tortoise species evolved to be tiny, such as Hinge-back Tortoises (*Kinixys* spp.). The Testudona lineage tortoises similarly decreased in size during their evolution. Tortoise body size, from gigantic to tiny, is apparently an evolutionarily labile (flexible) phenomenon that reflects local adaptation to conditions that are not limited to islands nor to particular continents. But the substantially lower average body size across tortoises today is a reflection of selection by humans for larger tortoises to consume (Chap. 9).

Continent by Continent

The combination of molecular and fossil data on tortoises makes for a complex story of continental origins. Before molecular data became available, it was easy to assume that each lineage of tortoise evolved on the continent where it is now found, with limited cross-continental colonization during a period of 50 or 60 million years. But molecular data have challenged that theory. The living tortoises today in Central and South America, Asia, and Europe all appear to be genetically affiliated with the Geochelona lineage, which suggests some complex migrations including across oceans. In describing continental origins, we acknowledge the many uncertainties that can only be resolved through further research.

Starting in Asia

Asia in the Paleocene (66–56 mya)

The current evidence suggests that tortoises (Testudinidae) arose in Asia during the Paleocene (~60 mya), and from there they dispersed to the other continents. Their spread from Asia was explosive, with North America and Europe colonized by the Eocene, and likely Africa as well. During the Paleocene, Europe and Asia were a single landmass that remained connected throughout the Cenozoic, although variously shaped and somewhat separated by an epicontinental sea. Thus tortoises could have spread into Europe overland or with a relatively short float.

Asia in the Eocene–Oligocene (56–23 mya)

Two species of Pleistocene *Manouria* (*M. yushensis* and *M. oyamai*) have been found in the Shanxi Province of China and the Ryukyu Islands of Japan, respectively, the only known members of the genus other than today's representatives (*M. emys* and *M. impressa*). Molecular data indicate that they have been there since the Late Eocene or Early Oligocene. Neither of these two species, belonging to the most primitive known tortoise genus, survived in eastern Asia beyond the Pleistocene.

Asia in the Miocene Upward (23 mya to present)

Fossil tortoises from tropical and eastern Asia are poorly known from the Miocene and onward. The major exception is *Megalochelys sivalensis* (formerly *Colossochelys atlas*). Owing to nomenclatural confusion and an overestimate of size, much has been written about it (Fig. 8.3). *Megalochelys* is represented by several Pleistocene fossil specimens, one from Punjab, India, and others from several of Indonesia's

Figure 8.3
Silhouettes of a dome-shelled living Galápagos Tortoise (*Chelonoidis niger*; *left*) and the even more giant extinct *Megalochelys sivalensis* (*right*). Sketch based on a silhouette illustration from Reptilis.net, accessed May 8, 2006

Sunda islands. The original fossil from Punjab is the largest, with a carapace length of 2 m (the height of a standard door). Its ancestors may be some of the earliest members of the group that now includes the large Leopard Tortoise (*Stigmochelys pardalis*).

Similarly, the two living *Geochelone* species (*G. elegans* and *G. platynota*) apparently had their origins from African ancestors. (Note that the term "Geochelone" was used for more than a century as the generic name for most fossil turtles from the Eocene to recent times, including species as diverse as the giant tortoises of the Galápagos and Aldabra Island. Now, it refers to just to the two living Asian species and their fossil predecessors.) The fossil history of true *Geochelone* is little known. They were potential residents of India when it crashed into the Asian mainland during the Eocene, although it seems unlikely that they were residents during its entire passage from the Southern to Northern Hemisphere that started more than 100 mya.

Into Europe

Until the 21st century, many tortoises from the European fossil record were grouped into the genus *Testudo* before deeper analyses sorted them into separate genera. Today, *Testudo* includes five living species (Chap. 7) plus a handful of extinct species.

Europe in the Eocene (56–34 mya)

Despite the molecular data showing tortoise origins in Asia, Europe boasts the most primitive-looking tortoise fossil remains found to date. A species dubbed *Fontainechelon cassouleti* (formerly *Achilemys*) from the Early Eocene of France (about 50 mya; range 55.8–48.6 mya) so far looks to be the most basal (earliest) of Europe's tortoises. Certain features, including for example a distinct pattern of neurals and no gular projection (Chap. 2), are presumed more primitive than those of *Hadri-*

anus majusculus, the species boasting the oldest fossil remains based on dating of its discovery locale (below and across to North America).

Other tortoise remains from the Eocene upward show various degrees of evolution from the primitive characteristics of *Fontainechelon*. For example, the body plan of *Pelorochelon soriana* found in Austria, France, Germany, and Spain about 40 mya (range 48.6–37.2 mya) looks a bit more advanced than that of *Fontainechelon*. Another one of the earlier European tortoises was *Cheirogaster maurini*, which may have been ancestral to subsequent Eurasian tortoises (a hypothesis that awaits more verification). *Cheirogaster* had low-domed and somewhat long carapaces. Their fossil remains, found only in France even though moderately common, are largely pieces of shells, thus providing little information on their evolutionary history, and *Cheirogaster* disappeared by the end of the Eocene (33.9 mya).

Europe in the Oligocene (34–23 mya)

Early Oligocene remains of a tortoise found in France were initially called *Testudo gigas*, thereby lumped into the genus *Testudo*, that encompassed all terrestrial turtles in Europe. But closer comparison of the fossils with other European tortoises showed sufficient differences to warrant its recognition as a distinct evolutionary group. Now dubbed *Taraschelon gigas*, it was large (almost 80 cm SCL) with a high shell, a deep notch on the nuchal plate (frontmost carapace bone) flanked by a wide pair of marginal (edges) scutes, and other unique characteristics. It differs from all known European and African tortoises in the absence of a cervical scute (see Fig. 2.4*A*) and other features.

Other tortoises (*Achilemys, Ergilemys, Hadrianus*) lived in Europe during the Oligocene, but all are considered problematic in terms of understanding relationships. Future molecular and morphological analysis will sort out the identity of these tortoise genera. Regardless, one species appears to have given rise to the Testudona lineage of today's small- to medium-sized tortoises (genera *Indotestudo*, *Malacochersus*, and *Testudo*), whereas another lineage would give rise to all other extant tortoises, their ancestors, and the now-extinct Mascarene *Cylindraspis* (except for the primitive *Gopherus* and *Manouria*). Much of this diversification occurred during the Oligocene and into the Miocene. Interpretation of European tortoise origins is complex, marked by migrations, radiation (rapid diversifications into more species), and extinctions.

Europe in the Miocene–Pleistocene (23–0.01 mya)

Miocene tortoise fossils are scarce in central Europe, found only in Austria, the Czech Republic, Germany, Hungary, and Switzerland. There are just two reports of

Miocene giant tortoise fossils that appear to be *Titanochelon* from the south Moravian region of what is now the Czech Republic (dating to 13.7–12.7 mya), a species that had a shell more than 100 cm long. More abundant Miocene fossils (estimated 20–11.6 mya) of a small tortoise have been found in Austria, the Czech Republic, and Germany; these may be the remains of *Testudo kalksburgensis* that reached a maximum size of about 25 cm long.

Still, the middle Miocene was a pivotal period for diversification of big European tortoises, which were nearly all *Titanochelon* with domed, somewhat elongate carapaces. Based on the Moravian (Czech Republic) fossils, European tortoises first attained large body sizes in the middle Miocene and were eventually to become abundant and diverse. *Titanochelon* is one of the largest known tortoises (from at least 1 m to nearly 2 m SCL) and lived in Europe through the Early Pleistocene; it consisted of eight or more species from Portugal and Spain to central Europe and Asia Minor. *Titanochelon* may have arisen from an earlier stock (which included the Oligocene *Taraschelon gigas*) as it shows no relationships to African tortoises or any other group of living tortoises.

Testudo hellenica is the oldest known member (Late Miocene, about 9 mya) of the lineage that today includes the Spur-thighed Tortoises (*Testudo graeca*). It had the hinged plastron of this clade and a carapace nearly 30 cm SCL. This recently (2020) discovered species was found in fossil beds that also contain remains of apelike animals, but there was no evidence that there was a predator-prey relationship. A small tortoise (20 cm SCL), *T. marmorum*, from the Late Miocene (8.7–7.75 mya) of Attica, Greece, also had the hinged plastron indicative of the Spur-thighed Tortoise lineage (and lived within its present range). It's presumably a descendent of *Protestudo* from the Middle Miocene of Turkey. The recently named *T. lohanica*—with front legs exceptionally well armored with osteoderms (bony skin deposits)—left behind abundant Late Miocene fossil remains in eastern Romania, making it the best-represented extinct *Testudo* tortoise.

By the Pleistocene, the Marginated Tortoise (*T. marginata*) lived across much of Greece, including on what is now the island of Crete. A subspecies (*T. m. cretensis*) was named from fossils found in the island's caves of tortoises larger than today's Marginated Tortoise. But analysis of their shapes relative to all the observed variation in living Marginated Tortoises suggests that the fossil tortoises may not merit a distinct subspecies. Regardless, there are no longer Marginated Tortoises on Crete, indicating a shrinking of their range after the Pleistocene, likely related to humans.

Fossil remains of a new species of tortoise, dubbed *Solitudo sicula*, found on the Italian island of Sicily, imply an animal of about 50–60 cm SCL, much bigger than the Hermann's Tortoises (*T. hermanni*) in the region today. Indeed, the skeletal remains do not look like Hermann's Tortoise, warranting a new name—*Solitudo sicula*—for the Sicilian species. Along with fossil tortoises from other Mediterranean islands such as Malta and Menorca, the remains suggest a non-*Testudo* tortoise lineage inhabiting the islands during the Pleistocene.

Into and Out of Africa

Africa in the Eocene (56–34 mya)

There is a dearth of identifiable tortoise fossils in Africa, making the timing of their colonization uncertain. Tortoises likely arrived in Africa over the connection to Asia (via what is now Arabia) by the Middle Eocene (46–34 mya), although it could have been as late as the Eocene-Oligocene boundary or as early as the Paleocene, based on molecular data. The colonists probably walked overland because the African plate was still connected to the Arabian plate until about 25 mya. Africa was to become the cradle for tortoise diversity, with one or several tortoises eventually giving rise to the diverse Geochelona lineage with more than 25 species.

Presently, the oldest known tortoise from Africa is the giant tortoise *Giganochersina ammon* from the Late Eocene of eastern Egypt (~35 mya). This tortoise is the first of many in this area—Afro-Arabia (North Africa from Morocco to Arabian Peninsula)—the region that has been considered the gateway for tortoise colonization of Africa. The habitat of the time, based on ecological inferences from the fossil record, was open savannah, in which high-domed tortoises tend to thrive. Colonization likely occurred earlier in the Eocene and, once tortoises became established, included movement in both directions. Evidence for their Eocene arrival includes the presence of several distinct lineages in Africa (judging from genomes) by the mid-Oligocene (~30 mya), combined with a genomic study indicating the close relationship of a group of South African, Madagascar, and Indian Ocean tortoise genera.

The five extinct species of Mascarene Islands giant tortoises (*Cylindraspis* spp.) were originally assumed to have dispersed eastward from Madagascar by sea. Recent analysis of DNA from their bones has shown otherwise, however. It tells of their arrival and evolution on the Mascarene Islands, which would become Reunion and Mauritius, approximately 40–45 mya during the Eocene (much earlier than the other Indian Ocean tortoises). The *Cylindraspis* tortoises likely came from southeastern Africa and represent a unique lineage with no survivors today.

Africa in the Oligocene (34–23 mya)

During the Oligocene, tortoises left a scarce fossil record in Afro-Arabia. A possible early Oligocene specimen of *Giganochersina ammon* has been reported, but not verified, from Oman. Even though tortoises obviously occurred in Africa during the Oligocene—given the earlier and later fossils bookending the era—none have been confirmed.

Africa in the Miocene (23–5.3 mya)

The Miocene fossil record for Africa is richer. Overall, sixteen species of tortoise appear in the Afro-Arabia fossil record, including four that are extinct, five that are still on Earth (extant), and seven more of uncertain identity. Colonization of Africa by tortoises was a pivotal event because evidence to date suggests that tortoises from Africa dispersed not only to the Indian Ocean islands, but also to both North and South America.

The only plausible explanation for tortoises colonizing the Indian Ocean islands is that they floated there. Currently, all evidence points to two colonization events from eastern Africa, one to the Mascarenes and another to Madagascar. Getting to the Mascarenes supposed a route north of Madagascar to the Réunion hotspot—the geologically active area that formed the Mascarene Islands. We presume that this was a single event, likely in the Miocene, and that the initial colonists gave rise to the ancestral *Cylindraspis*. The relationships of these tortoises to current African tortoise genera and species are not known.

The other colonization was to Madagascar and likely with more than one tortoise lineage successful arriving from east-central coastal Africa. For the giant tortoise (*Aldabrachelys*) lineage, speciation resulted in two species (*A. abrupta*, *A. grandieri*). Given the genetic similarity of *A. abrupta* and *A. gigantea*, it is likely that a pre-*abrupta* propagule floated from Madagascar to the granitic Seychelles islands and became *A. gigantea*, which then colonized Aldabra.

Africa in the Pliocene–Pleistocene (5.3–0.01 mya)

Once tortoises reached these islands, their populations fluctuated dramatically as sea levels rose and fell with changing climate. The repeated discovery of fossilized tortoise bones on Aldabra shows that the atoll was inundated by high seas several times in the past 170,000 years, including as recently as 15,000 years ago, and after each submergence it was recolonized by giant tortoises. Sampling of mitochondrial DNA sequences revealed no variation across three locations on two of the Aldabra Atoll's islands, which may be evidence of a colonization event by one or a few individual tortoises during the last interglacial period. Thus the current Aldabra population must stem from a relatively recent arrival, likely fewer than 10,000 years before the present.

The Late Pliocene witnessed extinctions of giant tortoises from the mainland as early humans appeared and began to use stone tools (~2.6 mya). The smaller tortoise species were presumably less sought-after for their meat. But fossilized tracks of tortoises recently discovered on the Cape south coast of South Africa indicate that very large tortoises (up to 50–60 cm long) lived in that area into the Pleistocene (~100,00 years ago). Whether they were either a species yet to be named or bigger individuals of the still-living Leopard Tortoises remains to be seen.

The Pleistocene glaciations isolated groups of tortoises, resulting in some speciation in Africa. For example, Home's Hinge-back (*Kinixys homeana*) and the Forest Hinge-back (*K. erosa*) likely evolved independently when glaciers caused their isolation in separate ice-free forests. Even with Hinge-back species, there is genetic differentiation that speaks to how populations have evolved independently; Home's Hinge-backs from Cameroon are genetically distinct from the same species in Ghana. Similarly, in southern Africa, Spur-thighed Tortoises show two distinct genetic clusters that are also reflected in differences in body size. The fluctuating environment of the Pleistocene may have contributing to differentiation of tortoises in Africa, followed by other ongoing factors that structure populations (Chap. 6).

Across to North America

North America in the Paleocene (66–56 mya)

The tortoises of North America have a rich fossil record beginning in the Late Paleocene. The oldest known tortoise fossils in the world were found in Paleocene-Ypresian strata (~53 mya) of western North America. *Hadrianus majusculus* was medium to large in size (at least to 53 cm CL) with a broad, arched, long carapace. It also had a big, wide head, plus sturdy front and hind legs with long, pointed claws. Other species of *Hadrianus* continued to appear through Eocene times (leading up to 33.9 mya) in North America. Even though its fossils are the oldest, *Hadrianus* had more advanced features than *Fontainechelon* found in France (see the Into Europe section).

Many extinct tortoise species from North America seem to be from the Geochelona lineage, although not closely related to South American tortoises. The implication, then, is that tortoises reached North America independently (apart from the South America dispersal event, discussed below). At the beginning of the Eocene (57 mya), North America and Europe had a broad land connection above the slowly widening Atlantic. More research is needed to clarify how the North American dispersals occurred.

North America in the Eocene (56–34 mya)

By the Late Eocene, North American tortoises were diversifying to include multiple genera. *Stylemys nebraskensis*, medium-sized tortoises (30–40 cm SL) with broad, low-domed carapaces, lived from the Eocene to Oligocene (fossils in Eurasia assigned to *Stylemys* are fragmentary and likely not *Stylemys*). Legitimate fossils of *Stylemys* are found in western North America deposits in California, Colorado, Montana, Nebraska, Oregon, and South Dakota, ranging into Saskatchewan, Canada. *Stylemys* was moderately abundant in the Late Eocene.

A more derived (advanced body plan) tortoise—*Hesperotestudo brontops*—ranged from South Dakota to what was the southernmost extent for Eocene tortoises in Oaxaca, Mexico. During the Eocene, *Oligopherus* became one of the major midsized (30–60 cm CL) groups of tortoises in North America. Both *Stylemys* and *Hesperotestudo* may have arisen and diverged early from the Geochelona lineage and would remain in North America through the Pleistocene.

North America in the Oligocene (34–23 mya)

The Oligocene witnessed a diversification of the *Hesperotestudo.* Their carapace shapes were variable, ranging from a high dome to a more elongated dome. Body sizes among the more than 20 species of *Hesperotestudo* to evolve also varied widely, ranging from the 30 cm SCL Wilson's Tortoise (*H. wilsoni*) to the Southeastern Giant Tortoise (*H. crassicutata*) measuring up to 1.5 m long (similar to Galápagos Tortoise size) that lived in southern North America. *Hesperotestudo* was a remarkably long-standing, geographically widespread, and common lineage of tortoises, but it notably did not penetrate South America. It occurred as far north as Saskatchewan (*H. exornata*, Early Oligocene) to (eventually) as far south as Florida (*H. alleni*, mid-Pliocene) and Texas (*H. johnstoni*, Early Pleistocene).

The Oligocene also saw the diversification of the *Oligopherus* lineage, which split into two lineages of Gopher Tortoises (*Gopherus* and *Xerobates*) by the early Oligocene. The fossil history of the Gopher Tortoises is better known than that of any other living tortoise genus. The oldest known species from fossils is *G. laticuneus,* which lived about 33 mya in what are now Colorado, South Dakota, and Wyoming. Up to 44 cm CL, *G. laticuneus* was intermediate in size to the smallest and largest members of its genus. The Early Oligocene witnessed a diversification of *Gopherus* into the half dozen species discovered to date that, like their living descendants, ranged in size from small to as large as *G. donlaloi* (CL > 50 cm). (*Note*: Although we are not recognizing *Xerobates* here, that name is accepted by some tortoise biologists and can be characterized by unique anatomical and behavioral characteristics visible in extant *G. agassizii, G. berlandieri, G. evgoodei*, and *G. morafkai.*)

Stylemys, in contrast, apparently suffered declining populations as climate cooled into the Oligocene; *S. nebraskensis* was replaced either by populations of a smaller body size or perhaps a new and smaller species of *Stylemys* still unnamed.

North America in the Miocene-Pleistocene (23–0.01 mya)

The diversity of North American tortoises peaked in the Miocene. The Berlandier's Tortoise (*G. berlandieri*) lineage appears to have arisen in the late Middle Miocene and diverged again in the Late Pliocene into *G. auffenbergi* and *G. berlandieri,* with only the latter surviving to the present. The larger, western desert tortoises—the Mojave Desert Tortoise (*G. agassizii*) and Sonoran Desert Tortoise (*G. morafkai*)—likely diverged in the Early Pliocene as climate dried and their respective desert areas (Sonora and Mojave) emerged.

The *Gopherus* lineage that includes today's Gopher Tortoises (*G. flavomarginatus*) and Bolson Tortoises (*G. polyphemus*) seemingly had greater diversification from the Early to Middle Miocene; five species have been discovered. Those include three fossil species from the Pliocene-Pleistocene: *G. depressus* from California, *G. hexagonatus* from Texas, and *G. donlaloi* from Mexico. The evolution of this group has largely occurred in the latitudinal band of 25°N–35°N, which made for semiarid landscapes with mild winters. The trait of burrowing well known in extant species of *Gopherus* (Chap. 6) may have arisen early, perhaps even in the *Oligopherus* ancestor, judging from the shape and bulkiness of its forearms and hands.

Looking across the Pleistocene, there were likely six distinct species of *Hesperotestudo,* although there was species turnover during glaciated periods, with new species emerging in interglacial periods. The newly discovered *H. harrisi* in New Mexico differs from its cousins on the Great Plains in some aspects of its jaws and palate, indicating that it was more of a grazer than a browser. The multiple presumed species of *Hesperotestudo* may eventually further sort out into a couple of distinct lineages that lived in different parts of North America.

During the mid-Pleistocene, *Hesperotestudo* colonized Bermuda and established a population that eventually evolved into *H. bermudae,* evidenced by a single adult specimen (27 cm CL) found on a limestone cliff face and a second specimen from an underwater cave. To reach Bermuda, a gravid tortoise would have had to float at least 1,500 km (in a straight line) from Florida, crossing the Gulf Stream and getting into a recirculation gyre (because the Gulf Stream passes west of Bermuda).

Many North American tortoise habitats would have become unsuitably cold during Pleistocene glaciations. Thus speciation can be attributed to differentiation as populations were confined to southern glacial refuges, evolved in isolation, and then reinvaded northern areas with the retreat of the glaciers.

Across to South America

Today's three modest-sized species of mainland South American tortoises are all classified as *Chelonoidis*, with just one giant species remaining, the Galápagos Giant

Tortoise (*Chelonoidis niger*). Yet the fossil record shows a diversity of *Chelonoidis* in South America from the Miocene through the Pleistocene, including giant species along the south-central portion of the continent in what is now Argentina. Piecing together the origins of South American tortoises has been difficult because the tortoise fossils are patchy and mostly concentrated in Argentina. The best molecular data so far is from a genomic study of a 1,000-year-old humerus (arm bone) from the now-extinct Bahamian Giant Tortoise (*C. alburyorum*). Its preservation in a water-filled sinkhole (absent oxygen) permitted a nearly complete genetic map of the relationships among all the extant and fossil species of *Chelonoidis*.

South America in the Eocene–Oligocene (56–23 mya)

Paleontologists have debated the origins of South American tortoises since the mid-1900s. Uncertainties about these origins arise from tortoise dispersal ability coupled with the human penchant for moving tortoises around as food or pets (Chap. 9). For example, tortoise remains on Sal Island, Cape Verde, off the coast of Africa, were initially thought to be from an extinct species dubbed *Geochelone atlantica*. DNA examination revealed them to be from a modern South American Red-footed Tortoise (*C. carbonaria*) that may have been brought to Sal Island by people running a local salt business during the 20th century.

Two competing 20th-century hypotheses purported that South American tortoises had arrived from North America or that they had come from Africa. During the relevant geologic time frame (35–25 mya), South America had long split from Africa, such that tortoises would have had to cross the Atlantic Ocean; there was no land bridge to North America at the time. Assuming that the *Chelonoidis* are indeed monophyletic (share the same common ancestor), it could have been just one cross-Atlantic dispersal, for example, of a gravid female (with eggs; Chap. 4) that floated across on currents.

Molecular and morphological (body shape) evidence both support this out-of-Africa hypothesis, showing *Chelonoidis* spp. to reside within the *Geochelona* lineage, which today includes mostly African tortoises. Exactly when tortoises dispersed from Africa to South America is uncertain, but it was during a period when the young Atlantic was narrower than today but still hundreds of kilometers wide and part of the global ocean circulation. A gravid female *Chelonoidis* likely made the crossing, landing on continental South America (rather than the Caribbean islands). To date, the oldest fossils found are Miocene, but molecular evidence indicates that a gravid female must have made the crossing sometime during the Late Eocene, diverging from African tortoises about 34–38 mya.

South America in the Miocene (23–5.3 mya)

The proposed trajectory has the original *Chelonoidis* diverging from Africa's Spurred Tortoise lineage (today's *Centrochelys sulcata*), floating to central South America, and beginning its radiation from there. To date, the oldest known South American tortoise fossils are from the medium-sized Gringo's Tortoise (*Chelonoidis gringorum*) from the Chubut Valley in Argentina's Patagonia, dating to the early Miocene. Here, genomic studies are in line with paleontological evidence. In most phylogenies, *C. gringorum* is ancestral to today's living Chaco Tortoises (*Chelonoidis chilensis*) and Galápagos Giant Tortoises; in some phylogenies, *C. gringorum* is even ancestral to all tortoises in the genus *Chelonoidis*.

South American tortoise colonists moved both northward and southward. The southern ones became the Chaco Tortoise that now lives east of the Andes Mountains in Argentina, Bolivia, and Paraguay. The northern ones that diverged from the shared *Chelonoides* ancestor an estimated 14–26 mya were later to seed the colonization of both the Galápagos Islands and islands of the Caribbean. This northern population diversified and eventually yielded the smaller Red-footed Tortoise (*C. carbonarius*) and Yellow-footed Tortoise (*C. denticulatus*). The now-extinct *C. hesterna* found in Colombia appears to be the last common ancestor of the Red-footed and Yellow-footed Tortoises. If so, they split around 13.5–11.8 mya. During the Miocene, however, smaller tortoises moved southward; their geographic distribution got displaced toward the lower, warmer latitudes as Miocene climate warmed.

But bigger tortoises appeared in South America during the Miocene too. Another set of remains found in the Chubut Valley, Patagonia, were from a large species (SCL up to 80 cm) dating to the mid-Miocene, about 15 mya. The so-called *C. meridiana* ranks as the oldest large *Chelonoidis* remains in South America, as well as the southernmost remains of tortoises this large in the world. The giant *C. gallardoi* later inhabited Patagonia in the Late Miocene. Farther north, at Quebrada Honda, Bolivia, lived other giant tortoises, representing the only Miocene turtle remains known from Bolivia. And Miocene fossils of a giant *Chelonoidis* sp. have also turned up in Colombia. Giant tortoises today are associated with tropical or subtropical climates, suggesting that at least the mountainous Patagonia locales were once warmer than they are now.

South America in the Pliocene–Pleistocene (5.3–0.1 mya)

One or more tortoises reached the Galápagos about 2 mya in the mid-Pleistocene. Molecular clock estimates show the Galápagos species starting to diverge from its mainland ancestor 1.5–2 mya, the islands having formed 4 mya. The diversity of

Galápagos species could have resulted from just one or two colonization events. It has been proposed that tortoises reached only a couple of islands from the mainland, including the oldest—San Cristóbal—and then spread through the Galápagos to eventually inhabit 10 islands by floating or rafting. The float distance today is a minimum straight line of 930 km from the mainland, but the islands were more distant then (and have been slowly moving closer to the mainland). Although it's possible that the original tortoises were small like today's Chaco Tortoises and evolved their giant size once on the islands, Pleistocene fossils from coastal Ecuador include fossil giant tortoises (see the Body Size Evolution section).

The *Chelonoidis* tortoises that colonized the Caribbean islands were likely from northeastern South America or southern Central America. Ancient DNA from the humerus bone of the extinct Bahamian Giant Tortoise showed its close relationship to Galápagos Tortoises and the mainland Chaco Tortoises. Tortoises colonized various Caribbean islands, with radiations beginning about 1.5 mya from the common ancestor that arrived by sea. A new tortoise, *C. dominicensis*, was described from fossils found in a cave on the island of Hispaniola, Dominican Republic (DR); a second tortoise, *C. marcanoi,* was recently named from dry caves in Pedernales Province, in the southwestern part of the DR; and there's a Cuban fossil, *C. cubensis*.

The Bahamian Island species started to diverge from their mainland ancestors about 1.5 mya, but the Bahamas could not have supported tortoises until at least 400,000 years ago because they were previously flooded during an interglacial warm period. Yet fossil tortoises from the Bahamian Islands sort out into two genetic groupings (subspecies) that must have diverged before they came to the Bahamas, which implies at least two dispersals. Even though multiple dispersals of tortoises to the islands sounds improbable, sea levels were much lower at the time, such that some islands had land connections. Tortoises may have floated from Cuba or Hispaniola, then spread out to the other islands. Fossil tortoises are known from 14 islands in the Lucayan Archipelago (Turks and Caicos plus Bahamas), including from some of the most remote islands.

Ocean currents from the mainland to the islands likely facilitated tortoise dispersal, with two possible routes (Antilles Current or the Caribbean-Yucatan Current). The closing of the Isthmus of Panama in the Late Miocene generated currents that start off the northeastern coast of South America and flow outward. Once tortoises reached the islands, geographic separation caused them to gradually diverge into distinct subspecies. That divergence is relatively recent, estimated at 0.09–0.59 mya for the Bahamas and 0.08–1.43 mya (Pleistocene) for the Galápagos.

The result of all this dispersal and differentiation was that by the end of the Pleistocene, diverse tortoises ranging from small to giant lived in South America and its neighboring islands. Many areas that today lack tortoises were inhabited by a diversity of species. Pleistocene giant tortoises—as well as smaller tortoises such as Chaco Tortoises or close, now-extinct cousins (e.g., *C. petrocellii*)—inhabited northeastern Argentina and the coastal Buenos Aires Province. Records of extinct *Chelonoidis* tortoises include those from the Bahamas, Barbados, Bermuda, Caicos, Cuba, Curacao, Grand Turk, Hispaniola, Mona, Navassa, and Sombrero Island. Ultimately, this whole radiation of Caribbean tortoises would be extirpated by human activities (Chap. 9).

Extinctions

It's difficult to estimate the total number of tortoise species that once inhabited Earth and are now extinct, partly because of the challenge of assigning fossils to species and partly because of the taxonomy itself. It's clear, however, that most tortoise species went extinct during the Pleistocene into the Holocene in tandem with human arrivals on each continent. In Asia, the two species of primitive *Manouria* did not make it beyond the Pleistocene. In Europe, the three Miocene *Testudo* (*T. hellenica, T. marmorum, T. antiqua*) all went extinct once hominins came onto the scene, but the genus is represented by five *Testudo spp.* living today. Both *Stylemys* and then *Hesperotestudo* would also go extinct by the end of the Pleistocene, leaving only the tortoise genus *Gopherus* in North America. All Caribbean island species went extinct at the boundary with the Holocene, 10,000 years ago after humans entered South America. Prior to human arrival, the island tortoises had no predators except crocodiles in some places, evidenced by bite marks on tortoise shells, and various predators on eggs and hatchlings.

The last giant tortoises on mainland Europe date to about 2 mya. They held out longer on islands, such as the Canary Islands, Malta, and Sicily. Still, the only remaining giant tortoise populations today are on the Galápagos Islands and Indian Ocean islands. The extinction of giant tortoises from mainlands likely caused particularly dramatic changes to their ecosystems, given the significant ecosystem engineering roles of tortoises. Giant tortoises are keystone species in ecosystems as vigorous plant grazers and browsers.

Tortoises may also have been the earliest seed dispersers, thus playing a critical role in the evolution of fruiting plants. Fossil shells of the approximately 30-million-year-old *Stylemys* tortoise in South Dakota contained fossilized Hackberry (*Celtis*) seeds. Necessarily, their loss would have had a profound effect on plant communities. Giant tortoises, with their larger home ranges, were likely especially efficient dispersers. Remains of the more recently extinct (4,200- to 1,200-year-old) Bahamian Giant Tortoise contained seeds from Wild Mastic and Satinleaf (*Sideroxylon foetidissimum* and *Chrysophyllum oliviforme*, respectively). In South America, giant tortoises dispersed "tunales" (*Opuntia* cacti of several species) and wild potatoes (*Prosopanche americana*). *Opuntia* patches now occur in areas from which tortoises are extinct, such as eastern Argentina, perhaps remnants of dispersal by the last Pleistocene tortoises.

So, tortoise evolution took place on the mosaic of shifting landmasses, associated islands, and dispersal corridors including land bridges and ocean currents. After having peaked in the Miocene, the diversity of North American tortoises (Testudinidae) declined to levels like we see today after human-caused rampant extinctions during the Pleistocene Epoch (ending 11,700 years ago) and the current Holocene Epoch (Chap. 9).

CHAPTER 9

Human Interactions I

Tortoise Decimation

In the world of turtles, tortoises are relatively late arrivals. Still, tortoises preceded humans on Earth by orders of magnitude. The oldest evidence for modern humans—*Homo sapiens*—is about 350,000 years ago. The genus *Homo* only goes back to 1.5 to 2.5 mya. For most of their history, tortoises have lived in a world free of humans, much less of modern humans.

When humans appeared and began to spread over the planet, conditions for land tortoises—especially the large and "giant" forms—changed dramatically for the worse. Tortoises are slow and easy to catch, with a body plan that had effectively repelled other predators for ~38 million years but did not defend them well against humans. Tortoises offer edible proteins and fats, which spurred extensive harvest.

Most if not all tortoise species suffer or have suffered from human exploitation, leading to their decimation and in some cases extinction. Of the 121 tortoise species that have overlapped on Earth with humans, 69 (or 57%) are now extinct. One of the most threatened groups of vertebrates, the tortoise species that remain are at risk of becoming extinct during the 21st century. Because many tortoises are long lived and slow growing, they take a long time to reach reproductive age. Juvenile mortality is high, so their populations rebound slowly or not at all if harvesting continues (Chap. 5).

Pleistocene (2,580,000 to 11,700 years ago)

Beginning in the Pleistocene, the abundance and diversity of wildlife changed dramatically as humans spread across the planet. The Pleistocene Overkill Hypothesis, which has been hotly debated since at least 1861, proposes that as humans colonized new areas during the Pleistocene, their efficient hunting methods resulted in the decimation and eventual extinction of many species, particularly the megafauna. From fossil evidence, paleontologists have established that in several regions, as the

Pleistocene proceeded, gigantic and large tortoises became relatively less common on continents but persistent on oceanic islands. This evidence of waning mainland populations is consistent with the argument that as humans dispersed over continents, tortoise species disappeared.

Indirect evidence reveals human consumption of tortoises during some of their earliest interactions, which is not surprising given that early human diets mirrored available resources in the regions they inhabited. Tortoises are easy to catch and nutritious in fats and protein. It has been estimated that a single Angulate Tortoise (*Chersina angulata*) provides about 3,332 kJ (796 kcal) of calories, more than a quarter of the daily energy needs for an active adult human. Where humans overlapped with tortoises, they ate them. Tortoise remains recovered from Pleistocene sites in both Europe and Africa bear evidence of human predation.

Europe

The earliest evidence of humans eating tortoises in Europe is the remains of Hermann's Tortoise (*Testudo hermanni*) found with bones of the first humans (about 1.1–1.2 mya) in a cave in northern Spain (Sima del Elefante). Cut marks on the underside of the carapaces were likely made by humans using tools to remove the viscera. Similar marks show up at more recent sites, such as Bolomar Cave in northern Spain (occupied by humans 300,000–150,000 mya), where Hermann's Tortoise shells have both cut and burn marks. Their leg bones bear human tooth marks. In mid-Pleistocene Europe, the harvest of tortoises continued. A cave in Mount Carmel, Israel (occupied 60–48,000 years ago), contained more than 2,000 bones of Spur-thighed Tortoises (*Testudo graeca*). Another cave site in Torres Novas, Portugal, occupied from 70,000 to 35,000 years ago contained 3,394 tortoise remains, some with burn or cut marks. Other sites in Italy, Spain, and the Near East also contain evidence of humans harvesting tortoise. Clearly, humans depended on tortoises for food throughout the Pleistocene.

Africa

Mid-Pleistocene (or "Stone Age") caves in South Africa contain evidence of what may be the earliest known human consumption of tortoises in the region. At a few sites, such as Pinnacle Cave, which contained more than 4,000 specimens of Angulate Tortoises and Parrot-beaked Tortoises (*Homopus areolatus*), cut and burn marks are apparent on their bones. Tortoise remains at Blombos Cave in the Western Cape, South Africa, indicate that during the thousands of years of human occupation (from 70 to 100,000 years ago), tortoises were a dietary staple.

North America

The arrival of humans in North America marked the beginning of the harvest of tortoises. The best evidence for human predation derives from southeastern North America and the Caribbean islands. Late Pleistocene remains of large tortoises from several Caribbean islands have signs of human consumption. The Little Salt Spring sinkhole in Florida contained the remains of a now-extinct giant land tortoise (*Hesperotestudo crassiscutata*) killed with a sharpened wooden stake and cooked in its shell; the remains date to 12,000 years ago. This sinkhole also contained other vertebrate remains and artifacts, confirming that it was used by Paleo-Indians as a natural trap.

Other evidence of prehistoric predation on tortoises in the Americas is indirect. Remains of now-extinct fossil tortoises occur at nearly 20 sites on Caribbean islands where there are no native tortoises today, but where prehistoric peoples are known to have lived tens of thousands of years ago. Harvest continued into the near present, for example, a 1,000-year-old shell of the now-extinct *Chelonoidis alburyorum* found in the Bahamas Sawmill Sink hole. Additional shells dated the extinction of this species to about 780 years ago, not long after humans settled the area. Although the fossil record of tortoises in South America is limited, there's evidence that species succumbed to a combination of frequent climate fluctuations and human exploitation around the Pleistocene-Holocene boundary.

By the Late Pleistocene, reductions in tortoise body sizes in cave middens at multiple locales indicate that human harvesting occurred often enough to change the age structure of tortoise populations with fewer large adults present. The Pleistocene witnessed a destructive relationship between humans and turtles as humans spread across Earth. Tortoise extinctions tracked human migrations across continents and onto island archipelagos. The Holocene ushered in a new Age of Man, which exacerbated the threats. The pronounced warming temperatures and the global retreat of glaciers led to the first sedentary human farming communities, which accelerated the alteration of habitats. Between the need to adapt to climate changes and inability to avoid human slaughter, tortoises lost out.

Holocene (11,650 to Present)

From the Pleistocene through today, about 70 tortoise species have become extinct. This number possibly represents more than half of the tortoise species that have ever existed. Compared to other turtles, which were also hard hit during the Age of Man, tortoises are the most vulnerable, as they are slow-moving terrestrial animals. At the dawn of the Holocene, Paleo-Indians already ate tortoises. For example, the Clovis People that colonized North America likely contributed to the extinction of *Hesperotestudo wilsoni* about 11,000 years ago. The larger Bolson's Tortoise (*Gopherus flavomarginatus*) ranged from Arizona eastward to Texas during the Pleistocene but was extirpated from Arizona, New Mexico, and Texas. North American aboriginal groups are known to have consumed *Gopherus* and likely contributed to its precipitous decline.

As the first Native Americans reached islands, a pattern of tortoise extinctions followed. For example, Abaco Island in the Bahamas was colonized around 950 years ago by the Lucayan Taíno People, who drove the endemic Bahamian Giant Tortoise (*Chelonoidis alburyorum*) to extinction in less than 200 years (780 years ago). Related giant tortoise subspecies (*C. a. keegani* and *C. a. sementis*, respectively) on the Turks and Caicos Islands were similarly extirpated through harvesting by the Taíno and Meillac Peoples.

Cultural Impacts

The Age of Man also saw the flourishing of cultural practices as humans formed stationary communities that brought new threats for tortoises. A female Natufian shaman (Israel) buried in a 12,000-year-old ceremonial grave was positioned with her head resting on a tortoise shell and her body flanked by the complete shells of more than 50 sacrificed Spur-thighed Tortoises. During the Stone Age, Spur-thighed Tortoises in northern Africa were made into personal body ornaments. A Native American burial in Joshua Tree National Park included 36 burned tortoise shoulder blades (scapulae), suggesting a ritualistic significance.

Turtles (including tortoises) have embodied healing, creation stories, and other cultural meanings for Indigenous peoples around the world, as evidenced by archaeological artifacts dating to at least 9,000 years ago. Between their cultural significance and the former abundance of tortoises in nature, their shells historically proved ideal for making ritual objects. The Greek "chelys" lyres featured in classic depictions were fashioned from tortoise shells strung with sheep gut that, based on their shapes, came from Marginated Tortoises (*Testudo marginata*). Shells from Mojave Desert Tortoises (*Gopherus agassizii*) were fashioned into ceremonial rattles by groups living around the Mojave Desert.

The cultural importance of tortoises to native desert tribes increased over time. Tortoise shells were used by Native Americans for functional, everyday items such as bowls, containers, scoops, and pottery tools, or for medicinal purposes. The Yavapai

People ground shells into powder as a medicine to treat stomach and urinary tract infections. In early Holocene Europe as well, tortoise parts were valued as commodities, as evidenced by carapace containers at sites in Romania dating to the 8th and 5th centuries BCE. Tortoise motifs on baskets, pottery, and rock art of both Mojave Desert tribes and the Mayans in Central America spoke to their cultural significance. A tortoise-like turtle was the central symbol in the Mayan calendar, with its scutes indicating segments of time.

Age of Sail

During the Age of Sail, tortoise populations faced new, greatly intensified harvest pressures from 16th- to 19th-century European pirates, explorers, and ultimately whalers. Although earlier contact with humans had been damaging, the levels of exploitation during the Age of Sail were downright devastating. Remaining populations of giant tortoises on islands were hit especially hard. Because tortoises are ectotherms (Chap. 3), they can sustain high population densities with limited resources, as demonstrated on these islands where they numbered in the thousands. But shortly after tortoises were found by seafarers, they became a staple food (meats, fats, and water-filled bladders), prized for their ability to withstand long ocean voyages.

Indian Ocean Islands

Prior to the Age of Sail, huge populations of *Cylindraspis* giant tortoises (five distinct species) lived on the Mascarene Islands east of Madagascar (Mauritius, Réunion, and Rodrigues plus smaller remnants of volcanoes). French Explorer François Leguat, the first to publish descriptions of these giant tortoises on Rodrigues Island, wrote in 1691 of "two or three thousand of them in a Flock; so that one may go above a hundred Paces on their Backs . . . without setting Foot to the Ground." Clearly, their abundance was staggering, believed to be in the hundreds of thousands on Rodrigues alone. Recently, a historic nesting site was found on Rodrigues Island, complete with egg clutches.

When the Island of Mauritius was discovered by the Dutch and later settled, giant tortoise populations declined rapidly and were rare as early as 1671. Tortoises on the islands of Réunion and Rodrigues were not far behind in suffering a similar fate. As ships visited the Mascarene Islands, they collected thousands of large tortoises (*Cylindraspis* spp.). A tortoise-hunting station was established on Rodrigues to supply the growing colonies of people in the islands, with an estimated 280,000 tortoises harvested from Rodrigues from just 1732–1771. Both giant tortoise species and a near-giant living on Madagascar when humans arrived—the Madagascar Giant Tortoise (*Aldabrachelys abrupta*), Grandidier's Giant Tortoise (*A. grandidieri*), and Bour Tortoise (*Astrochelys rogerbouri*)—were extinct by the early 15th century, likely earlier.

Other Indian Ocean islands suffered the same intensive exploitation, including the Seychelles, discovered on January 19, 1609, by a British ship exploring for the East India Company. It was not a day later that the seamen loaded as "many land tortells as they could well carrie." The captain's journal reported the giant tortoises tasted as good as fresh beef but were repellant to some men because of how ugly the tortoises looked before cooking. The Seychelle tortoise populations had a reprieve from exploitation during the early 18th century, but after settlement in 1768, tortoises became the main export product. Ships carrying 200–250 tortoises each regularly shuttled turtles from Seychelles to Mauritius. Ships anchored in the harbor depended on tortoise meals to such an extent that from 1784 to 1786, 3,000 giant tortoises were supplied to naval vessels. By 1803, tortoise populations were substantially depleted, and the Seychelles was importing tortoises from Aldabra to supply local needs and for export.

The Giant Tortoises (*Aldabrachelys gigantea*) of the Aldabra Atoll experienced a similar exploitation, but thanks to the absence of good anchorage and the roughed terrain with dense shrub coverage, not all tortoises could be found and removed. Their safety was not ensured because during the Cold War in the 20th century, the US military wanted a base in the Indian Ocean to track Soviet maritime activity in this region. Aldabra offered a low, flat, remote oceanic island (60 sq. mi.) with a firm limestone base on which a large airfield could be built diagonally across the widest part of the atoll to handle large military planes. The British government was willing to lease the island for joint usage as a naval and air base. There was no consideration given to the plant and animal inhabitants, particularly vast numbers of roosting and nesting sea birds. Fortunately, before the transaction occurred, the impending agreement came to the attention of British ecological community, which along with journalists, activists, and scientists of the Royal Society of London managed to head off the military development. Intense biological studies followed and continue. The United Nations Educational, Scientific, and Cultural Organization (UNESCO) now recognizes Aldabra as a World Heritage Site.

Galápagos Islands

The exploitation of Galápagos Giant Tortoises (*Chelonoidis niger*) off the coast of Ecuador did not begin immediately after their discovery by European seafarers. The islands were found by Spaniards early in the 16th century. The first known ship to enter the area was in 1535 when Bishop of Panama Fry Tomas de Berlanga drifted off course on his way to Peru, but it would be a century before people revisited the islands and harvest of tortoises began. First, it was a trickle of English privateers during the 1600s, although many 18th-century explorers just bypassed the Galápagos, thinking the stark terrain had little to offer. Whaling changed things dramatically, however.

By the late 18th century, the British identified the Galápagos as a key area for capturing sperm whales. American whalers followed suit. A growing demand for whale oil drove Pacific whaling. Whalers reportedly harvested many thousands of tortoises. In less than 40 years—from 1831 to 1868—American ship logs show that whalers collected more than 13,000 giant tortoises from the Galápagos Islands. The tortoises were kept on board for weeks or months with no food or water; thus they served as the first "canned food." By midcentury the United States had a 700-ship whaling fleet based out of New England, whose sailing logs reported single ships taking as many as 900 tortoises on board (Fig. 9.1) for the sailors' food supply.

Because male giant tortoises are larger than females, the exploitation was biased toward females that could more easily be carried. David Porter, a US Navy officer who visited the islands in 1812 to defend American whaling ships from the British, wrote that of 14 tons of tortoises taken aboard his ship, "only three were males." The skewing of harvest toward females eliminated the production of young and the abil-

Figure 9.1
Galápagos Tortoises were taken on ships for food or commerce, such as these on the Chilean *Valdivia* in 1903. Photograph by Carl Chun, Freshwater and Marine Image Bank

ity of the populations to recover from these depredations (Chap. 6). As the tortoises became rarer, seamen had to search farther inland for them. By 1858, an American captain wrote that "In order to get them we had to go high up in the mountains, as that seems to be their roaming ground."

The rush for whale oil, and accompanying consumption of tortoises, continued until commercial petroleum became available in the 19th century as an alternative to whale oil. By the turn of the century, Galápagos Tortoise populations numbered in the hundreds; there had once been thousands. Whaling ships had harvested an estimated 100,000 tortoises from the Galápagos Islands with more casualties from changes wrought by human uses, such as settlements. Direct harvest (Fig. 9.2), coupled with introductions of dogs, pigs, and rats that consumed eggs and damaged habitats, resulted in the extinction of at least three subspecies of Galápagos Tortoises.

Some individual tortoises ended up in scientific collections for research, supplying much of the data now used to understand their history. For example, in 1834, ship commodore John Downs donated two Galápagos Giant Tortoises to the Bos-

Figure 9.2

Even with reduced populations at the beginning of the 20th century, Galápagos Tortoises were being exported and offered for sale, as in this 1902 California market. Photograph from *The Wasp* (January–December 1902)

ton Society of Natural History. Considered a living curiosity, they were alive when they arrived, but most did not live long. Their remains are tagged specimens at the Harvard Museum of Comparative Zoology. The fascination with giant tortoises and their resilience drove scientifically minded people to send live tortoises to the gardens and museums of Europe and the United States where their survival rates were poor, both owing to insufficient knowledge of their needs and to unsuitable climates, such as that of Massachusetts.

The main outcome of giant tortoise exploitation on islands around the globe was a spate of extinctions. The Galápagos Islands originally harbored 15 subspecies, according to fossil records and molecular data. Three of those went extinct, and today six others are critically endangered. In the Indian Ocean, once home to 17 species of giant tortoises, all but one had gone extinct by 1840. Only the one species on Aldabra Atoll remains.

Today

Today, the main threats to tortoises are habitat change and direct exploitation. Tortoises are heavily harvested for food, traditional medicine, cultural purposes, and even as pets. The relative severity of the threats varies by species and environmental conditions. The biggest threat to Geometric Tortoises (*Psammobates geometricus*), for example, is habitat loss and degradation in South Africa, whereas for Ploughshare Tortoises (*Astrochelys yniphora*), it's direct exploitation for trade in pets and food. At this point in the history of our relationships with tortoises, continuing exploitation is directly linked to impending extinctions.

Food

The period of intense exploitation of large giant tortoises from their island refuges was over not long after the Age of Sail ended. Still, numerous Galápagos Tortoises and even a few Aldabra tortoises are lost to poaching every year.

For other tortoise species, humans continue to directly harvest eggs and adult tortoises for local, subsistence consumption. For example, harvest of the various species of Hinge-back Tortoises (*Kinixys* spp.) in central Africa is ongoing. Tortoises are a major item in the bushmeat trade. In South America, some traditional cultures regularly consume tortoises, such as the Mebêngôkre (Kayapó) and the Waiwai in central Brazil, for whom monkeys, agoutis, and tortoises are dietary staples; or the Kuikuro, for whom turtle meat is a backup when fishing is inadequate. For the Kayapó, tortoises are traditionally the food of choice for ceremonial feasting. Groups of men trek into the forest and return with enough tortoises to roast for the whole community. In other cultures, tortoises may not be specifically targeted by hunters, but they are nevertheless captured when encountered. Yellow-foot Tortoises (*Chelonoidis denticulatus*) in the Brazilian Amazon are incidentally captured, kept in pens until their digestive system has been cleansed with a vegetable diet (no carrion), then slaughtered for celebratory events such as Easter or Christmas.

Historically, tortoise meat was a local staple, for example, Radiated Tortoise meat eaten in Madagascar at Easter and Christmas celebrations. As tortoise meat has become more popular, however, a larger-scale problem has emerged—the exploitation of medium-sized tortoises from Africa, Asia, and South America to supply the wild meat market, sometimes with disturbingly high levels of take. Monitoring of trash dumps for a couple hundred days in a southwestern city in Madagascar documented the dumping of almost 2,000 shells from Radiated Tortoises (*Astrochelys radiata*), which are critically endangered. Most of those were headed offshore, presumably for export of bushmeat, despite Madagascar's wildlife legislation that prohibits any harvest of the species. A decline in the Radiated Tortoise population of nearly 50% in the decade leading up to 2012 was largely attributed to poaching and harvesting for meat.

Even in regions of the world where religious prohibitions have prevented people from eating turtle flesh (such as Muslims of Africa's Sahel region), famine and political instability have resulted in consumption of tortoise meat (including African Spurred Tortoises, *Centrochelys sulcata*). Wars or other crisis tend to drive up exploitation of natural resources and override controls. Civil war in Liberia has caused pressure on Home's Hinge-back Tortoises (*Kinixys homeana*) as soldiers or people escaping fighting in forested areas eat them. Bending of rules is also common. Although the Antandroy People of southern Madagascar have a taboo on eating tortoise that was maintained even through severe droughts and famines, some Tandroy collected Radiated Tortoises to sell to other people (non-Tandroy) in the region that did not have prohibitions against tortoise meat.

Medicine

Because humans covet tortoise longevity, tortoise body parts have become fetish or pharmaceutical items in traditional medicine. In Brazil, Yellow-foot Tortoise meat and fats are used to treat rheumatism. In Morocco, tortoise carapaces have traditionally been used for hand tremors, insomnia, anxiety, and curing victims of witchcraft, not to mention taming quarrelsome husbands. Traditional West African cultures harbor beliefs about magical healing properties of reptiles as "zootherapeutics." An inventory of reptiles traded in markets specialized in traditional medicines of Togo, West Africa, revealed African Spurred Tortoises and three species of Hinge-backs either traded as entire dried specimens or shells only. One of the most commonly traded was the Western Hinge-back Tortoise (*Kinixys nogueyi*), a species listed in Appendix II of the Convention on International Trade in Endangered Species of Wild Fauna and Flora (CITES). Similarly, a survey of markets in Mozambique found *Kinixys* for sale as a species in demand by traditional healers. There is evidence that tortoise presence in local markets has declined, which likely simply reflects scarcity.

Some tortoises appearing in markets today may come from ranches, such as African Spurred Tortoises that are captive bred in Togo. Yet many wild tortoise populations have a biased age structure that reveals repeated loss of certain age classes from the population. For example, as early as 2001, Moorish Spur-thighed Tortoises had populations skewed toward juveniles in west-central Morocco. In contrast, this species in the Maamora Forest had more age-balanced populations in protected areas relative to unprotected. Spur-thighed Tortoises sold in Marrakesh markets were mostly small ones, likely reflecting harvest from nonprotected areas. Because tortoises require a long time to mature, these skewed populations will require many tortoise generations until the populations recover balanced age distributions (Chap. 4).

Pets and Trafficking

Tortoises are popular pets. The global pet trade has become one of the main current threats to their survival. The attractive coloration of many species, their convenient sizes and low maintenance in captivity, and ironically the lure of their rarity drive the market. Tortoise popularity is clearly a double-edged sword, illustrated by a study of French children that showed their dual desire to protect, but also to own, tortoises.

Depending on the region, people make pets of local, native species. Spur-thighed Tortoises, for example, are common pets in the cities of Morocco, living in more than half of 480 households surveyed in a study of Rabat and neighborhood villages. Many of the pet tortoises are juveniles reportedly collected from the wild by their "owners." As urban areas become broader, people encounter wild tortoises, magnifying the problem of local collection. In a study of exurban sprawl in southeastern Spain, half of new residents keeping tortoises had gotten them from the wild with the intent to protect them from perceived threats.

Although people plucking tortoises from the wild and taking them home is problematic, the captures for export and international trade are likely orders of magnitude more devastating. For example, Home's Hinge-backs that could be sold for export became an attractive target for Ghana locals that had previously made money by collecting and selling snails. The harvest rapidly decreased Home's Hinge-backs in the wild. On the receiving end of the pet trade, in Texas alone from 1995 to 2000, 16 species of tortoise were imported and exported as pets or zoo animals. International wildlife trade is well established, and its networks are sometimes linked to other activities, including the arms and drug trade. This drives a cycle of intense, damaging tortoise exploitation in places where it is least sustainable.

From the largest international trade hubs for tortoises in African and Asian countries, many tortoises go to Europe or the Americas for sale as pets. A 2016 review of international turtle trade pointed to the United States as the top importer, with China and Hong Kong not far behind. After 2005, Asian exports of wild turtles declined, but North American exports increased. From 1990 to 2010, of the 10 families of turtles subject to trade (Fig. 9.3), tortoises were the most heavily exploited group, with an estimated 750,000 individuals crossing international borders. Marketing tortoises as pets is not new.

Even when CITES imposes export quotas, tortoises continue to be traded under the table as pets, with quotas regularly exceeded. For example, all tortoises native to Thailand are protected under the 1992 Wild Animal Reservation and Protection Act. Yet Bangkok's huge weekend market (Chatuchak) frequently features tortoises

Figure 9.3
Wild-caught Sulawesi Tortoises (*Indotestudo forestenii*) in a Sulawesi animal trade's pen awaiting export, most likely for the pet-trade market. Photograph by Cris Hagen

for sale, even those banned from trade. In just three days of observation, the sale of nine species of tortoise, including many of Madagascar's endangered Radiated Tortoises, were documented at Chatuchak. Penalties and fines are apparently too low to discourage the continuing lucrative pet trade.

Bangkok's market, where one can easily obtain information from vendors about how to smuggle tortoises across borders, serves as a hub for trade in tortoises that are not native to Thailand. From 2000 to 2005, Thailand reported an import of 1880 Indian Star Tortoises (*Geochelone elegans*), purportedly brought in through Kazakhstan and Lebanon, the latter not a party to CITES. From there, tortoises are bought by dealers who take them back over borders to Japan, Malaysia, Singapore, or elsewhere to be sold as pets. Additional monitoring to detect illegal imports and exports is clearly needed.

Illegal wildlife trade was historically transacted purely through physical markets and exchanges (Fig. 9.3). But emerging platforms on the Internet have become additional conduits for negotiations of tortoise sales. Much of the online trade occurs in plain sight but is not yet regulated. Despite its protection under both CITES and Madagascar law, the critically endangered Ploughshare Tortoise appeared for sale on multiple online retail platforms during 2015, with prices ranging from $509 to $47,000. Similarly, native, protected Spur-thighed Tortoises and Steppe Tortoises (*Testudo horsfieldii*) were advertised for sale on an Iranian online marketplace (Sheypoor) as recently as August 2021, the ads mostly coming from provinces within the native ranges of those species.

The online tortoise trade tends to at least start on the domestic level rather than internationally because of language barriers. For example, on a Hong Kong Internet turtle forum that includes opportunities to purchase turtles, posts are in traditional Chinese. A one-year study (2017–2018) identified more than 29,000 turtles for sale, of which most were still sold in physical markets but augmented by social media and other Internet platforms. Once purchased domestically by traders, tortoises head to international markets.

Even the giant tortoise species are vulnerable. An illegal shipment of 185 juvenile Galápagos Tortoises thought to be from nests on Isla Santa Cruz was confiscated in March 2021 after an attempted illegal exportation. Each hatchling was wrapped in plastic, and 35 died from the experience.

Habitat Change

In the wild, tortoise populations decline as their habitats are converted to other uses. Human activities typically fragment favored tortoises' habitats into remnants surrounded by unsuitable habitat. Because tortoises rely on specific habitat features and the continuity of these features, there are fewer and smaller habitats available to tortoises, thus concentrating them into smaller areas (Chap. 6). For example, the fragmented populations of Gopher Tortoises (*Gopherus polyphemus*) in Mississippi have lower hatching success of eggs and more reproductive abnormalities in adults. Their reduced genetic diversity—from historic severe reductions in their populations (bottlenecks)—adds insult to injury as populations become more isolated. A population with low genetic diversity, and limited opportunity to increase it through mating with more distantly related individuals, can gradually accumulate harmful mutations.

Even where suitable habitats remain, human activities often degrade them for tortoises. Cattle alter natural habitats in numerous ways that make then unsuitable for many wildlife species. In the semiarid Sahel, the African Spurred Tortoises populations are smaller in the presence of cattle. These tortoises prefer areas with taller vegetation along stream banks, including seasonal streams ("kori"), and are only active during the wet season. Cattle compete with tortoises as plant grazers in the short term and cause desertification in the long term, creating conditions too dry and barren for tortoises. Mojave Desert Tortoise burrows and juvenile tortoises are trampled by grazing sheep. In South Africa, landscape changes for human activities have eliminated more than 90% of Geometric Tortoise habitat.

Many tortoise species occupy habitats that are relatively open and naturally maintained by fire. As natural fires are suppressed, these habitats may fill in with more shrubs and trees that ultimately reduce habitat suitability for tortoises. Southern US Longleaf Pine (*Pinus palustris*) habitats that are favored by Gopher Tortoises, for example, are frequently surrounded by human development, which suppresses natural fire regimes. The increase in canopy cover that results, often from an infusion of fast-growing invasives such as Cogon Grass (*Imperata cylindrica*), does not provide suitable tortoise conditions and forage. Studies show that Gopher Tortoises abandon burrows when canopy closure increases beyond the natural, fire-maintained level.

Tortoises that naturally inhabit forested areas, such as tropical forests, lose access to food and other resources when forests are modified or reduced. Madagascar's Spider Tortoises (*Pyxis arachnoides*) live in dry, coastal forests that are losing natural vegetation at a rate of more than 1% per year to small-scale agriculture, fuel wood harvesting, and production of charcoal. The species is critically endangered, with its relic populations increasingly restricted to smaller areas and its population declining at ~1.4% per year. Its survival rests on alleviating the poverty that drives forest conversion. Such small-scale habitat conversions occur throughout the distribution of the world's tortoises.

In a few instances, tortoise populations adjust to habitat change. Spur-thighed Tortoises in the Souss Valley in Morocco have been losing their habitat to wood collection, grazing, and agriculture. They now survive in the edges of agricultural lands and have changed their diet to the introduced Prickly Pear plant. The fleshy, spined cactus provides food as well as shelter and sites for estivating (Chap. 3). In Florida, one Gopher Tortoise population had higher adult survivorship and body size in a restored area than in their natural sandhill habitat. So, there are a few cases where tortoises successfully adapt to new and novel habitats.

Roads and Fences

Despite the protection of a hard shell, tortoises are regularly squashed on roads throughout the world by motor vehicles. Crushing by vehicles—both on and off roads—is the main source of Mojave Desert Tortoise mortality in the American West (Fig. 9.4). Mojave Desert Tortoise home range sizes decrease the closer they are to roads, perhaps an adjustment to higher disturbance levels at the costs of reduced

Figure 9.4
In their search for water, Mojave Desert Tortoises (*Gopherus agassizii*) put themselves in danger by drinking from roadside puddles. Photograph by Tyler Green, National Park Service

access to the resources that naturally dictate home range size (Chap. 6). Because tortoises are most active during the reproductive season, adults may be particularly vulnerable to becoming roadkill. The loss of gravid females is a significant impact to any population. The tendency for tortoises to move slowly across roads and pause in the face of danger also makes them particularly vulnerable.

While fencing roadsides to prevent tortoises from entering roadways is a strategy to reduce direct mortality (Chap. 10), it divides habitat in ways that may have population-level drawbacks for tortoises. Like habitat conversion, fences create barriers that effectively divide habitat into smaller pieces. The Karoo of South Africa, home to several species of tortoise, has been crisscrossed with fences beginning in the 1880s as a means of livestock management as well as other linear barriers such as railroads, canals, and more recently solar facilities.

As tortoises attempt to cross these barriers during their normal movements, they're exposed to new hazards. Angulate Tortoises in the Karoo get wedged into mesh fences and become stuck there as they try to push through. Mojave Desert Tortoises moving along fences and roadways experience direct sunlight, which raises their body temperatures to near their physiological limits. Roads, power lines, and other human habitat modifications have increased the density of Common Ravens (*Corvus corvax*), which although native are voracious predators of tortoise eggs and hatchlings. Dirt roads in Palm Springs, California, have brought more mammal predators such as bobcats into proximity with tortoise burrows.

Another problem with roads, fences, or other barriers that tortoises are unable to cross is reduced genetic connectivity and diversity of populations. Fences may restrict seasonal movements for mate seeking and nesting, as well as limit the longer-term dispersal that promotes gene flow. Analysis of relatedness of Mojave Desert Tortoises on the Nevada-California border shows several genetic clusters in valleys separated by mountains but with genetic interchange. Across a railway and highway, gene flow is unnaturally reduced, potentially compromising resilience of the species to localized disturbances where historically tortoises could recolonize from a neighboring population.

All fence types generate tortoise mortality, but electrified fences are a particular hazard because tortoises don't recognize and avoid them. As early as 1994, electrified fences became a source of mortality for tortoises. Along 8.4 km of fencing in the Thomas Baines Natura Reserve, South Africa, 50 Leopard Tortoises (*Stigmochelys pardalis*) were found dead, mostly young adults. Thousands of tortoises are reported killed annually by electrocution from fences. A recent study of Leopard Tortoises highlighted continued negative changes in population structure that will eventually lead to extinction of populations near electric fences.

Contaminants

Human activities produce high volumes of chemical and other wastes that enter natural ecosystems, especially from agricultural activities. For example, in southern Greece, a population of Hermann's Tortoises was exposed herbicide spraying from 1975 to 1984. Mortality rates of those tortoises relative to noncontaminated populations increased by ~34%. Health effects ranged from swollen eyes and nasal membranes to immobility, with juveniles being most affected. The ultimate outcome was local extirpation of the population. Even today, with better regulation of agricultural chemicals, tortoises continue to suffer from contaminants. A 2016 study of Europe's reptiles found ongoing high risks of exposures of both Hermann's Tortoises and Spur-thighed Tortoises to pesticides. The southern parts of the European Union, richer in agriculture (vineyards and olive groves), are areas of particular concern. Consider that many regions of the world have less rigorous regulation of agricultural chemicals than the EU. A study of more than 100 protected areas in Latin America showed the continued intrusion of chemical contaminants.

Generally, as animals that have long life spans and process lots of vegetation, turtles can accumulate large quantities of chemical contaminants. When females nest, some contaminants transfer to their offspring via the eggs. How chemical contamination builds and spreads in tortoise populations needs further study. A study of heavy-metal concentrations in Mojave Desert Tortoises in California revealed that although soils contained elevated levels of arsenic, lead, and thorium, they were not accumulating in tortoises' bloodstreams.

Diseases

Disease is also a factor in early mortality of tortoises as tortoise populations become more confined and stressed. Tortoises carry a range of pathogens including viruses, bacteria, and fungi, which may be introduced via ticks or other parasites. Speckled Dwarf Tortoises (*Chersobius signatus*) in the Karoo biome of South Africa were found to host 34 bacterial species, a substantial load of fungi and yeasts, cysts from protozoan parasites (*Coccidia*), and nematode worm eggs. Galápagos Giant Tortoises carry both adenovirus and herpesvirus. Many of these organisms and viruses may be commensal (not harming the tortoises), and stressors like drought or other climate conditions can cause deleterious shifts in tortoises' microbial flora and sicken the individuals.

As habitat degrades, tortoise populations are squeezed into smaller or less suitable areas, which increases disease transmission rates. One of the most devastating tortoise diseases is caused by the bacteria *Mycoplasma agassizii*, named for its effect on Mojave Desert Tortoise populations. An infectious pathogen, it causes an upper respiratory tract disease that was first reported in the late 1970s in California in captive tortoises and likely was spread into wild populations by their release. By the mid-1980s, the disease was widespread in Mojave Desert populations. Soon thereafter, the disease appeared in Gopher Tortoise populations. Outbreaks can overwhelm a population. On Sanibel Island, Florida, an outbreak killed nearly half of the adult Gopher Tortoises. Mojave Desert Tortoise populations were halved from 2004 to 2013 from this disease.

Tortoise mycoplasmosis disease results from both the bacteria *M. agassizii* and its cousin *M. testudineum*, and perhaps also from other *Mycoplasma* spp. These debilitating respiratory infections are now known from wild tortoises of several species in multiple countries, as well as in captive Spur-thighed and Hermann's Tortoises in the United Kingdom. When it doesn't directly kill, mycoplasmosis nevertheless alters hormone levels and other aspects of tortoise physiology, which disrupts reproductive behaviors. The stuffy (mucus-filled) nose can also make it harder for tortoises to find food by blocking their smell receptors, leading to nutritional problems.

The infection, at least in Mojave Desert Tortoises, spreads between individuals through direct contact. Individuals may contract the infection when they're engaged in courtship or in combat. The infection can affect any age or sex of tortoise, but hatchlings are particularly vulnerable, with a high risk of death within six weeks of being infected. Fortunately, female tortoises who have been exposed to the pathogen convey helpful antibodies to their offspring through the eggs, providing immunization to their hatchlings. Drought, introduction of captive tortoises, and exposure to heavy metals can precipitate outbreaks.

Climate Change

Earth's ongoing climate change will affect tortoise populations just as it affects all living things. Tortoises living in dry climates depend on the relatively rare winter rains, during which free water is available for them to drink and rehydrate. A study of 18 desert reptiles found that Sonoran Desert Tortoises (*Gopherus morafkai*) were among the most vulnerable to climate changes based on ranks using NatureServe's Climate Change Vulnerability Index (CCVI).

Physiological Stress

As ectotherms, tortoise body temperatures fluctuate with ambient temperatures; climate has direct metabolic consequences (Chap. 3). If ambient temperatures regularly exceed their critical thermal maxima, tortoises will die.

Even when not regularly exceeding tortoise tolerance limits, temperatures play a key role in tortoise life history cycles from beginning to end. Gopher Tortoise populations in the southeastern United States, for example, show a gradient of growth rates, ages to maturity, and reproductive outputs. Generally, tortoises grow more slowly, mature later, and lay fewer egg clutches in the cooler north relative to the warmer south. Thus changes in seasonal temperatures can disrupt these life history features. Peak air temperatures, for example, raise levels of the estradiol sex hormone in female Bolson Tortoises. The rate of tortoise egg development and incubation time are tuned to temperature and humidity. Aspects of offspring development are vulnerable to disruption, such as hatchling sex ratios in tortoises, which are temperature dependent (Chap. 4).

A study of Hermann's tortoises found that the survival of juvenile tortoises is closely tied to the amount of winter rainfall. As seasonal conditions become increasingly dryer from climate change, the mortality of young tortoises will increase, quickly altering population structure and the survival of tortoise populations. During the 1997–2002 drought, Mojave Desert Tortoises had a higher mortality. Tortoise corpses showed signs of dehydration and starvation. As the summer of 2022 demonstrated, future droughts with extremely hot temperatures will render most Mojave Desert habitats unlivable for tortoises.

Studies demonstrate that the body condition of tortoises, measured by blood indicators, is lowest at the end of the dry season, owing to reduction of available water and food. Research on the Karoo Dwarf Tortoise (*Chersobius boulengeri*) revealed a strong correlation between growth rates of juveniles and adults with rainfall. In an extremely dry year, even shell volumes shrank, likely owing to the shrinkage of the viscera and bone resorption. Climate change models predict that as conditions become dryer, the range of this species will contract. Their already precarious population—the only remaining one for this species—has limited resilience, given an age distribution indicative of decline (Chap. 5).

Species Responses

Owing to different ecologies, different tortoise species are more or less vulnerable to climate change. Among North American tortoises, the arid-land species are more dependent on cyclic rainfall than their Gopher Tortoise cousin. Models of Mojave Desert Tortoise habitat change as climate warms and precipitation decreases predict a reduction of suitable habitat by more than 88% at moderate predictions of climate warming by 2°C and rainfall increasing by 50 mm. Gopher Tortoises live in the humid subtropics where rain is less limiting; they can count on eating plants with high water content year-round. But climate change alters natural events such as storms and fires, which shape Gopher Tortoise habitats that are maintained by fire. Will the future climate maintain the landscape necessary for the survival of Gopher Tortoise populations? It is hard to predict, since extreme weather events alter habitats and affect survival. Hurricane Matthew (2016) and Hurricane Irma (2017) hit northeast Florida. Both flooded low-lying habitats and drove tortoises to higher ground, as evidenced by the increase in more upland burrow density.

Tortoise populations in the American West are already feeling the effects of climate change. Mojave Desert Tortoises in Joshua Tree National Monument were devastated by persistent drought from 1997 to 2002 due to death by dehydration or by coyotes turning to tortoises as available prey when mammals became scarcer. Low-lying desert areas inhabited by tortoises are expected to become increasingly unsuitable as climate continues to warm. For example, researchers modeled shifts in suitable Mojave Desert Tortoise habitat due to changing climate by the end of the 21st century in California and found a reduction in livable area by 55%. Despite their resilience to short-term changes in conditions, tortoise populations succumb to sustained periods of inhospitable temperatures as well as water and food scarcities.

Ecosystem Responses

Shifts in climate are also accompanied by shifts in predator-prey dynamics. Even when tortoise predators are native—evolved alongside tortoises in many cases—climate change can alter their relationships. For example, Speckled Dwarf Tortoises in South Africa declined by more than 60% from 2000 to 2015, despite minimal changes in habitat during that period. Increased predation by Pied Crows (*Corvus albus*) on hatchlings was implicated in the decline. Even though Pied Crows are native to South Africa, their populations have dramatically increased in response to climate warming in the Karoo and Fynbos shrubland habitats of Speckled Tortoises.

In the future, as climate causes shifts in tortoise habitats, protections must be in place so tortoise populations can respond. Modeling the future availability of habitat for African Spurred Tortoises predicted an increase in land with suitable climate, but a decrease in quality of tortoise resources. Mapping current tortoise distributions against such modeling predictions should inform management. For example, across the entire range of Pancake Tortoises, only about 22% of their habitats are within protected areas. Models show that as climate warms, about 65%–85% of the climatically suitable habitat will lie outside of protected areas. Currently, the African Protected Areas Network offers some protection to Pancake Tortoises, but it will become increasingly less relevant as climate warms.

Tortoise populations that live on islands or near coasts face the particular challenge of losing habitat to rising sea levels. The next century is predicted to reduce

Aldabra Giant Tortoise populations by 45%–60% as their habitat is inundated by rising ocean waters.

Adapt or Migrate

How tortoise populations around the world weather climate changes will depend on their capacities to adapt in place or migrate into climatically suitable areas. Given the rate at which climate is changing, adaptation via natural selection is unlikely. Rather, tortoise populations may adapt through plasticity in aspects of their life history, such as timing of reproduction and egg deposition, or foraging patterns to capitalize on shifting resources. At least in a population of giant Galápagos Tortoises, longer-term average conditions (implying a certain amount of hardwired behavior), not immediate environmental conditions, will shape their movement patterns. The timing of their annual migrations from lowlands to uplands and back are not optimized to a particular season's conditions but are rather optimized to past conditions, implying a limited ability adjust to a rapidly changing climate.

Interestingly, climate predictions for the Galápagos Islands archipelago forecast warmer, wetter ecosystems, which may benefit tortoises through increased plant forage. More food availability leads to more growth, earlier age of maturity, and thus a longer reproductive life. Conversely, the higher rainfall may flood tortoise nests in those populations that nest in low lying areas. How species and populations of tortoises are affected by climate change will depend on the complex interplay between habitat requirements, plasticity, and resilience. Coastal Gopher Tortoises facing increased hurricane frequency can only compress their spatial use by so much, as they are increasingly squeezed between the ocean and upland development. Some other tortoise species may have more room to move, although human populations around the world have co-opted much of their habitats.

CHAPTER 10

Human Interactions II
Tortoise Conservation

Chapter 9 discussed the destructive treatment of tortoises by humans in the past. Such maltreatment still occurs, although there have been improvements and an increasing concern for the future survival of tortoises. We humans now dominate the planet Earth. Tortoises and all other life require our consideration of their needs and actions to ensure their continuity as species. Recent decades have witnessed individuals, groups, and governments stepping forward to help protect, rescue, and otherwise try to ensure a future for tortoises.

Pre-20th Century

Before the 20th century, most people worldwide gave little thought to tortoises other than as a food and commercial resource, with the exception of certain Native peoples. Notwithstanding the tortoise harvest by some Native American tribes, other Indigenous cultures revered tortoises as spirit animals and thus took measures to protect them. The Mojave Indians, for example, considered the Mojave Desert Tortoise (*Gopherus agassizii*) to be sacred and prohibited tortoise consumption. Ancient Mojave pottery and desert rocks bear depictions of tortoises, which were also honored through legends and songs as symbols for eternal life and the foundation of the Earth.

These traditional values would shape tortoise conservation decades later. For example, when part of the Mojave Desert called Ward Valley was selected by the state of California during the 1990s as a site for burying radioactive nuclear waste, the five tribes native to the area (Cahuilla, Chemehuevi, Cocopah, Mojave, and Quechan) joined forces with environmental groups, lawyers, scientists, and other residents to successfully protest the location. The court action resulted in the establishment of 6.5 million acres of habitat for the Mojave Desert Tortoise.

For the most part, however, when Native Americans were forced onto reservations near the end of the 19th century, they lost agency over ancestral lands, and protecting North American tortoises fell to legislative bodies and conservationists. Similar patterns played out on other continents.

Conservation Legislation

In the late 19th century, Albert Günther of the British Museum realized that the continuing harvest of the giant insular tortoises would cause their extinction. His studies of the anatomy and systematics of giant tortoises in the context of their dwindling populations brought an urgency to his research. After he published a monograph demonstrating the uniqueness of the Pacific Galápagos (*Chelonoidis nigra*) and the Aldabra Giant Tortoises (*Aldabrachelys gigantea*), he began writing letters to colleagues concerning their impending extinction. Although Günther's letters generated attention and resulted in small steps toward their conservation, no significant measures were taken to protect the giants or their smaller cousins. Conservation activities were focused elsewhere, such as stemming commercial hunting of wildlife that had brought the Passenger Pigeon and the American Bison, among others, to near extinction.

The Lacey Act

The Lacey Act of 1900 was the first major legislation aimed at stopping commercial exploitation of wildlife. Inspired by the observation that game animals were becoming scarcer, the act intended to prevent hunters from poaching game animals and escaping with them across state lines. Right on its heels came the conservation movement launched during the presidency (1901–1909) of outdoorsman and hunter Theodore Roosevelt, which led to the protection of federal lands as national forests, reserves, game preserves, refuges, and parks. Although tortoises were not a direct beneficiary of Roosevelt's conservation programs, they benefited from the overall landscape protection.

The Endangered Species Act

Following this burst of activity, interest in conservation declined during the Great Depression and the world wars. Not until the 1962 publication of Rachel Carson's *Silent Spring* was conservation again highlighted, this time under the rubric of environmental protection. Carson's evidence for the effects of the synthetic pesticide DDT (dichlorodiphenyltrichloroethane) on birds called attention to declining numbers of other wildlife and the continuing loss of natural areas with the ever-expanding human population. Although the Lacey Act had regulated commercial hunting of wildlife, it had not given the US Department of Interior's Fish and Wildlife Service (USFWS) authority to adequately protect endangered species. In 1966, Congress passed the Endangered Species Preservation Act, which provided both funding and authorization to protect, conserve, and restore native species of fish and wildlife.

When the act was later amended and strengthened to include a list of endangered species, the first list of foreign taxa in June 1970 included the Galápagos Tortoise and Madagascar's Radiated Tortoise (*Astrochelys radiata*). The domestic list of October 1970 did not include tortoises because the focus was on more charismatic mammals and birds. In 1972–1973, President Nixon requested that the US Congress pass comprehensive endangered species legislation, then signed the Endangered Species Act (ESA), which set the stage for listing of endangered tortoises. The first to be listed was the Bolson Tortoise (*Gopherus flavomarginata*) in 1979, followed by the Gopher Tortoise (*G. polyphemus*) populations west of the Tombigbee River in 1987. Mojave Desert Tortoises were not listed until 2014. So, official recognition of imperiled tortoise population was slow coming. Today, the ESA, which seeks to first protect species and then try to recover their populations to a level that no longer requires protection, remains one of the world's most significant environmental legislations.

Beyond the United States

Another major conservation event—with global reach—was CITES (Convention on International Trade in Endangered Species of Wild Fauna and Flora), which was proposed in 1962 at the annual meeting of the IUCN (International Union for the Conservation of Nature) of about 184 nations. The final document opened for signatures in 1973, was ratified in 1975, and included all tortoises by 1977. CITES offers protections to tortoises from trade and commercial exploitation worldwide, although unfortunately it's only as effective as its enforcement, which has proved to be challenging. The Convention on Biological Diversity (CBD) came into force in 1993. Signed and ratified by eventually nearly 200 countries, the CBD serves as a negotiating body for strategic actions and plans to protect biodiversity. Although its focus is on the broad maintenance of the natural services that biodiverse ecosystems supply, tortoises in theory should benefit from such ecosystem protections in areas they inhabit.

Many countries besides the United States have national or regional legislation that protects tortoises (Fig. 10.1). For example, in the most tortoise-diverse region—South Africa (Chap. 7)—tortoises are protected under the Cape Nature Conservation ordinance of 1974 that regulates hunting, the establishment of provincial

Figure 10.1
An illegal shipment of Egyptian Tortoises (*Testudo kleinmanni*)—of various ages, including adults—intercepted by Italian customs officials. These individuals may have come from a hobbyist's breeding compound, but trafficking these endangered species would still be illegal.
Photograph by Stefano Alcini, Wikimedia CC-BY-SA-3.0

nature preserves, and the appointment of nature conservation officers. Pertaining to the Cape of Good Hope, it specifies that all tortoises and turtles are protected. The more recent Biodiversity Act 10 of 2004 covers the entire Republic of South Africa and mandates a permit for any activities that may affect a listed species. Similarly, Asian countries that boast rich tortoise diversity have national laws that encompass tortoises: Thailand's 1992 Wildlife Preservation and Protection Act, Malaysia's 2010 Wildlife Conservation Act, and the 2017 Law of the People's Republic of China on the Protection of Wildlife.

Still, the prevalence of illegal trading and the challenges of enforcement compromise the effectiveness of these laws in protecting turtles. In the ongoing onslaughts from poaching and habitat degradation, the fate of many imperiled tortoise species rests in conservation programs specifically designed to protect and restore imperiled species.

Conservation Programs around the World

Few tortoises and tortoise species would survive today if it were not for the individuals and organizations—both government and nongovernmental—that advocate for the protection of tortoises and their habitats worldwide. These protection efforts have encouraged much of the research on tortoise biology that informed this book. Conservation programs take several approaches to support tortoise populations, including research and monitoring, habitat protection and restoration, and captive rearing and reintroduction. These activities often occur in combination. Although not comprehensive, the following eleven case studies demonstrate how some of the most imperiled tortoise species have been managed and, in some cases, brought back from the brink of extinction.

Galápagos Tortoises

Giant tortoises were some of the first to gain protection, beginning in the mid-20th century. Although by then human harvest had waned, invasive species such as rats, cats, dogs, and goats were decimating Galápagos Tortoise populations by eating eggs and hatchlings, and by reducing habitat quality. When Austrian ethologist Irenäus Eibl-Eibesfeldt visited the islands in the mid-1950s, he was shocked at the devastation of giant tortoises and shared his concerns in lectures and the book *Galápagos: Die Arche Noah im Pazifik*. His concern was echoed by other biologists, and an international group petitioned the Ecuadorian government for permission to initiate conservation activities on the island. In 1959, on the centennial of the publication of Darwin's *On the Origin of Species*, the Charles Darwin Foundation of the Galápagos was founded, and Galápagos National Park was established.

By 1965, the foundation, in collaboration with the Galápagos National Park Directorate, built a research station on Isla Santa Cruz (Fig. 10.2). Of the original 14 Galápagos Tortoise populations, just 11 remained, and some of them were on the brink of extinction. The tortoise conservation program has saved several subspecies from extinction through a combination of captive rearing, reintroductions to islands, protection of nests, and control of invasive animals. Captive rearing was first focused on the population of Isla Pinzón, where fewer than 200 adults remained. Wild eggs transferred to the Isla Santa Cruz station yielded 20 hatchlings that were eventually reintroduced to Isla Pinzón. Gradually, tortoises from other islands, including the 14 remaining tortoises on Isla Española, were brought to the program for breeding, rearing, and reintroduction.

Goals for reintroduction to Isla Española included not only establishing a viable population, but also restoring the functions of giant tortoises as ecosystem engineers;

Figure 10.2
Entrance sign to the Charles Darwin Research Station's breeding area for tortoises and other species. Photograph by Ken Dodd

they maintain open savanna habitats by grazing and tromping around. Galápagos Tortoises used to number in the thousands on Isla Española (and other islands), thus playing a keystone role in the ecology of these ecosystems (Chap. 6). Introductions to Isla Española began in 1975, and 35 years later yielded a self-sustaining population of some 800 tortoises showing a positive growth trajectory. As the population transitions from newly introduced to approaching carrying capacity (maximum number of tortoises that can be supported by the environment), management focus has shifted from ensuring their initial survival to maintaining the resources they rely on, such as tree cacti.

This initiative to save Galápagos Tortoises is considered one of the world's most successful conservation breeding programs. The diverse human interventions it required were illuminated in the path of a giant tortoise from the time it was collected off the islands in the 1930s to its role in today's conservation program. "Diego" was taken from Isla Española to the San Diego Zoo for a breeding colony intended to ward off extinction of the species. There he remained for 40 years, but as breeding faltered, he was returned in 1977 to the Galápagos National Park and Charles Darwin Research Station. There, he mated with multiple females to sire almost 800 offspring that were reintroduced across the Galápagos Islands. Diego is just one example of a long-lived Galápagos Tortoise that—with human assistance—helped salvage the species.

Genetic studies are essential to guiding captive rearing approaches to bolster genetic diversity in repatriated populations. For example, analysis of Galápagos Tortoises on Isla Isabela revealed three genetic types that corresponded to different ecological niches on the island. Ecological divergence after the island was colonized by tortoises likely resulted in these clusters. Although they currently do not constitute a distinct species, knowing the genetic patterning has stimulated discussion about how to manage these populations to maintain their respective genetic diversity. The Isla Floreana subspecies (*C. n. niger*) has been extinct since the mid-1800s, but parts of its genome live on in the descendants of individuals transported to Isabela Island that hybridized with its Volcán Wolf tortoises.

Linda Cayot, who spent her career as a PhD biologist working to support Galápagos Tortoises, wrote: "A research and management program with more than 40 years of success is only possible with the passion and dedication of many scientists, park wardens, consultants, student, and volunteers." The giant tortoise program on the Galápagos Islands earned its success from capturing the remaining tortoises and breeding them in captivity in "Galapagueros," reintroducing them to the wild and eradicating the introduced predators that compromised their habitat quality. The need for monitoring and management of these populations is perpetual.

Aldabra Tortoises

In the 1960s, the proposed construction of a military base threatened the integrity of the entire Aldabra Atoll. Fortunately, the Royal Society of London stopped the proposal, followed immediately by a surge of biological investigations on the atoll and surrounding waters. In 1970, a biological station was established to observe the island's flora and fauna, managed by the Seychelles Islands Foundation. By that time, native giant tortoises of the Indian Ocean were found in the wild only on Aldabra, having been extirpated from other Seychelles islands by humans and introduced predators. Aldabra Giant Tortoises are a keystone species; their grazing and browsing establish and maintain plant communities that support them and other species. Although Aldabra's tortoises were harvested, the lack of good anchorage and the roughness of the terrain allowed a residual population to survive. When regular harvesting ceased, the tortoises reproduce naturally and rapidly increased in numbers.

Although Indian Ocean giants were extinct elsewhere in the 19th–20th century, people had begun to translocate Aldabra Tortoises between islands. They were moved around so much without recordkeeping that sources of captive tortoises on islands are mostly unknown. The Seychelles Islands Foundation brought some order to the chaos, with reintroductions recorded from the 1970s onward. From 1978 to 1982, about 250 tortoises were reintroduced to Curieuse Island in several batches with the intent of starting a colony for tourists to enjoy. By 1990, the Curieuse population was reduced to half from theft, premature death from predation, and low reproductive rates. Still, it became a thriving tourist attraction, perhaps drawing damaging attention away from the wild tortoise populations on Aldabra.

Dozens of Aldabra Tortoises have also been introduced to other Seychelles islands since the 1970s (e.g., Bird, Cousine, Desroches, Farquhar, Grande Soeur, Moyenne), often from groups caged and penned on the larger islands. Despite their poor survivorship in some cases, introduced Aldabra Tortoises have helped restore ecological processes such as seed dispersal and soil turnover that supports native plants and insects (Chap. 6). In conjunction with removal of invasive animals and introductions of extirpated birds and plant species, tortoises have helped reverse the damage to Cousine Island in particular, effecting a sustained improvement in native habitats.

Aldabra Tortoises have also been introduced to the Mascarene islands of Ile aux Aigrettes, Mauritius, Rodrigues, and Round. Although *Aldabrachelys gigantea* was never native to those islands, it is considered a suitable "taxon substitute" for the *Cylindraspis* tortoises that once lived there. Introducing a tortoise that has a similar size, habitat use, and life history may restore some of the ecological roles played by the extinct *Cylindraspis* tortoises that vanished in the mid-1800s. Those roles, as described above, include maintaining vegetation by grazing and seed dispersal. Introductions have led to breeding populations of giant tortoises in the wild on Seychelles islands that had not harbored tortoises for more than a century.

The keystone role that tortoises play as herbivores has made them candidates for introductions to islands where no giant tortoises have previously lived. Basically, they would serve as mowing machines to keep invasive plants in check, as well as agents of native seed dispersal. Even on islands that have never had megaherbivores (grazing animals weighing more than 1,000 pounds), invasive plants overrun native vegetation. Rewilding islands with giant tortoises might help restore native vegetation where the original herbivores (such as the giant flightless Moa-Nalo duck in Kaua'i, Hawai'i) no longer survive.

Ploughshare Tortoises

Smaller tortoise species did not gain international attention as early as the giants, nor have they been entirely ignored. Madagascar had lost its giants a century or more ago, but by the middle of the 20th century, its other species were also being decimated by harvest for food, predation by feral bush pigs, and collection for the pet trade. By the end of the 20th century, the Ploughshare Tortoise (*yniphora*, also known as "Angonoka")—Madagascar's largest extant species—was the most threatened of Madagascar's tortoises, with around 400 individuals remaining. Ploughshare Tortoises are particularly vulnerable because they're restricted to a tiny natural range of bamboo scrub around a bay in northwestern Madagascar (Chap. 7).

In 1986, Gerald Durrell's Jersey Wildlife Trust in collaboration with the Malagasy Department of Water and Forests established Project Angonoka (Plate 24) to conserve and manage Ploughshare Tortoises. The program included using tortoises confiscated from illegal captivity to start a breeding colony in Madagascar's Ankarafantsika National Park, supported by an auxiliary colony at the Jersey Zoo on Jersey Island, Channel Islands. The first trial release of captive-reared juveniles was successful and set the stage for repopulating this species into the wild. Even with a setback, the theft of more than 75 individuals in 1996, the breeding program managed to increase the wild population to about 1,000, although they have since declined in the face of continuing threats from poaching for the pet trade and severe fires.

Hermann's Tortoises

In southern France during the 1960s, a biological investigation of Hermann's Tortoise (*Testudo hermanni*) showed that a protected reserve was needed to sustain its populations. In 1986, naturalists Bernard Devaux, David Stubbs, and Ian Swingland founded the association SOPTOM, short for Station d'Observation et de Protection des Tortues et leurs Milieux in le Var, for their protection. Once the resident herd reached about 150 in improvised enclosures around Devaux's home, he and his neighbors opened the first Tortoise Village—a sanctuary that operates as a protected area and education center (Fig. 10.3). Over the next couple of decades, other Tortoise Villages opened in Corsica (France) as well as in Senegal and Madagascar for the protection of other species. Thus SOPTOM grew its mission to include international conservation and research on tortoises in Eurasia and Africa.

Radiated Tortoises

Radiated Tortoises are endemic to Madagascar. Once common across the southern-southwestern region of the country, their presence today is greatly reduced from poaching, habitat changes, and international pet trade (Chaps. 7 and 9).

In the late 1990s, the Tortoise Village in southern France (see the Hermann's Tortoises case study) inspired the founding of a similar program in southwest Madagascar. There, Radiated Tortoises that are confiscated from illegal trade are integrated into a breeding program. In 2009, for example, 300 Radiated Tortoises arrived at the Itafy Tortoise Village. Another 400 were added in 2010. Suitable sites for reintroduction are scarce, but the 250,000-ha Lake Tsimanampetsotsa Reserve proved

Figure 10.3
Tortoise village sign in southern France, the home of SOPTOM's first breeding facility for Hermann's Tortoises (*Testudo hermanni*). Photograph by Ken Dodd

to be a viable location. Reintroductions occur through a collaboration between the Tortoise Village and public authorities (Madagascar National Parks, Departement de l'Environnement des Eaux et Forêts). These captive tortoises at Itafy also bring in money from ecotourism. Staff conduct awareness and outreach programs with the goal of stopping tortoise poaching and highlighting the fact that tortoises are more valuable alive than dead.

In 1999, Radiated Tortoises were also collected and moved to a Wildlife Conservation Society breeding management facility on St. Catherine's Island, Georgia, and later to the Turtle Conservancy's facility in Southern California for continued captive management. These captive populations allow researchers to study aspects of Radiated Tortoise reproduction that inform the restoration of wild populations. A challenge of breeding these and other tortoise species is generating a suitable sex ratio, given that their sex determination is temperature sensitive (Chap. 4). By tightly regulating incubation temperatures, the Turtle Conservancy program generated both males and females, with a female-biased sex ratio that tends to be favored in conservation programs given the importance of females in sustaining recruitment (of hatchlings).

Today, one of the last remaining strongholds of Radiated Tortoises is the Lavavolo Classified Forest in southwestern Madagascar. It is associated with a community of farmers and fishermen who subsist in a dry region with scarce resources. In 2008, a collaboration between multiple regional and international organizations (Madagascar Biodiversity Partnership, Omaha's Henry Doorly Zoo, Radiated Tortoises Species Survival Plan, Turtle Survival Alliance, and Turtle Conservation Fund) targeted the Lavavolo region for Radiated Tortoise conservation, based on molecular and field data. The multidisciplinary approach included research, education, community engagement, and technological transfer. By combining community development (e.g., providing regular access to water) with tortoise conservation, the program launched a tortoise nursery, a network of conservation educators, the buy-in of community elders, school-based tortoise programs, and habitat restoration plans. Radiated Tortoises gained local advocates while helping elevate the living conditions of the humans that share their habitats.

Hinge-back Tortoises

Hinge-back Tortoises (*Kinixys* spp.) have been poorly studied relative to other African species, delaying action regarding their conservation. Recognizing that all eight *Kinixys* species were likely declining (although not all are officially listed as threatened species), ecologists drafted a 2014 blueprint for their conservation. The first assurance colony (a captive colony to ensure survival of the gene pool if the wild population becomes extinct) was established for Home's Hinge-backs (*Kinixys homeana*) with about 100 starter tortoises. Subsequently, successful captive breeding has occurred in the United States and Europe for several Hinge-back species.

Captive maintenance is considered a stop-gap measure until pressures on wild populations and loss of Hinge-back habitat abate sufficiently to warrant reintroductions. Reintroductions are rare for these species. In Cote d'Ivoire, however, specimens provided by the Abidjan Zoo were released and monitored through a cooperative effort of the Office Ivoirien des Parcs et Reserves, African Chelonian Institute, and Rare Species Conservatory Foundation. Better wild protections—such as strict export quotas, trade monitoring, and restrictions on their sale in markets—are needed to secure the sustainability of any reintroduced populations of Hinge-backs.

Wild harvest for bushmeat, medicines, and international pet export continues within the African ranges of Hinge-back species. In some cases, such as in Ghana, where surveys for Home's Hinge-back have been conducted since 2010, the number of tortoises harvested per day has decreased, most likely owing to their increasing scarcity. Home's Hinge-back is safe from harvest in some African communities because it is venerated as a holy animal in the Niger Delta. Nigeria and Togo have national laws preventing harvest or international export of Home's Hinge-back; however, enforcement is weak. In Ghana, a community bylaw declared by a chief has achieved traditional protection of Home's Hinge-back in a few communities. More such local conservation efforts, with support to develop accompanying ecotourism or other income generating activities, may be the only way to stem extinctions of the wild populations of *Kinixys.*

Karoo Dwarf Tortoises

Conservation plans for dwarf tortoise species are limited by insufficient knowledge; they're small and secretive, spending much of their time well hidden in their habitat. Data collected on the only known Karoo Dwarf Tortoise (*Chersobius boulengeri*) population, endemic to South Africa, revealed a population in decline with few hatchlings or juveniles (Chap. 9). Predation by crows is implicated as the main cause for the decline of the Karoo Dwarf Tortoises and related species, such as the Speckled Dwarf Tortoise (*C. signatus*). Biologists recommend measures to keep crows away, such as removing perches (old telephone poles and windmills), eliminating food sources (landfills, livestock carcasses, and road kills), and reducing open water sources, and restoring habitats degraded by overgrazing. Conservation is particularly difficult for species like these that live in remote, poorly studied habitats.

Burmese Star Tortoises

Endemic to the dry region of Myanmar, Burmese Star Tortoises nearly went extinct owing to human consumption, use as traditional Chinese medicine, and the international pet trade. By the early 2000s, its status as extinct in the wild initiated a captive breeding program to save the species. In 2004, about 175 Burmese Star Tortoises (*Geochelone platynota*) confiscated from illegal wildlife trafficking were used to establish three breeding colonies that were managed by the Myanmar Ministry of Environment, Conservation and Forestry in collaboration with private conservation organizations (Wildlife Conservation Society and Turtle Survival Alliance). The rationale was that ensuring future survival of species imperiled in the wild calls for at least three groups of at least 50 tortoises each, with an approximately equal sex ratio to hedge against natural disasters or diseases.

Other breeding colonies were established over time, including one started with 150 tortoises pens at a Buddhist monastery. Each tortoise was marked with icons that represented religious taboos to discourage poachers. Captive rearing of this species proved to be relatively straightforward, and breeding was successful with at least 14,000 Burmese Star Tortoises hatched in breeding colonies. Beginning in 2014, Burmese Star Tortoises were released back into wildlife reserves. Although this species still faces pressures from poaching and habitat degradation, including in the protected areas, it has been brought back from the brink.

Gopher Tortoises

Within the United States, numerous organizations protect tortoises and their landscapes. At the federal level, all are within the Department of Interior and include the Bureau of Land Management and the Fish and Wildlife Service. Aside from providing protection for tortoises in the lands they manage, many staff members are involved in tortoise research. The wildlife departments of the states harboring populations of North American tortoises, such as the Florida Fish and Wildlife Conservation Commission, also are involved in tortoise habitat protection and have staff members studying tortoise biology. Importantly, these federal and state agencies provide grants and contracts to biologists in colleges and universities for their staff and students to investigate tortoise biology. Some other countries provide similar support to tortoise research.

Gopher Tortoises are endemic to sandy habitats in the southeastern coastal plain of the United States that face development pressures. They inhabit upland areas with longleaf pine that are particularly attractive to timbering interests and urban development ever since colonization by Europeans. By the late 1970s, their imperiled status was increasingly apparent. Biologists began rescuing Gopher Tortoises from sites slated for development and releasing them elsewhere, such as to St. Catherine's Island off the coast of Georgia and the Aiken Gopher Tortoise Heritage Preserve in South Carolina. At those release sites, the now-homeless "waif" tortoises fared as well as other wild tortoises already living there. Translocating waif tortoises may be a tool for augmenting other declining populations.

Beginning in 2014, Gopher Tortoises were also headstarted and released. For example, 145 tortoises were released in several groups into the Yuchi Wildlife Management Area in Georgia. The release area had previously become unsuitable for tortoises because of farming and tortoise harvest, pressures that had abated since being purchased by the state to become protected land. Survival of the released juvenile Gopher Tortoises was highly variable. Deaths were mostly from predation by fire ants or mammals such as raccoons. Staggering the releases over space and time ensured that at least some of the cohorts had higher survival rates.

Public-private partnerships are also pivotal in salvaging Gopher Tortoise populations. For example, the US Department of Agriculture Natural Resources Conservation Service in Mississippi provides financial and technical assistance to landowners to improve habitat for Gopher Tortoises. The Nature Conservancy and Mississippi Military Department established a captive rearing (headstart) program for Gopher Tortoises with the help of multiple stakeholder organizations: US Fish and Wildlife Service; US Forest Service; Mississippi Natural Heritage Program; and Mississippi Department of Wildlife, Fisheries, and Parks. Military lands managed by the Department of Defense are strongholds for Gopher Tortoises, which inhabit 28 military installations in the southeastern United States (Plate 11).

Mojave Desert Tortoise

In addition to the governmental investments, a variety of other US organizations have been founded to protect tortoises. The oldest organization is the Desert Tortoise Preserve Committee Inc., founded and continuously active since 1974. This nonprofit volunteer organization is dedicated to the conservation of the desert tortoises and other rare and endangered species living in the Mojave and western Sonoran Deserts. In 1980, recognizing the rapid declines in Mojave Desert Tortoise populations, the Bureau of Land Management established a 39.5-acre Desert Tortoise Research Natural Area that the committee oversees.

Augmentation of declining Mojave Desert Tortoise populations by headstarting (shepherding individuals from hatchling to a certain size/stage) is also part of their recovery plan. Beginning in 1990, multiple headstarting programs began raising young desert tortoises and, in the process, learning about their early life stage needs. Females were brought from the wild to headstart enclosures to lay eggs, then returned to their natural habitat. The results were mixed, with high hatching rates but poor hatchling survival, which in the earliest programs was a result of overcrowding and drought. A later program that started at Edwards Air Force Base in 2003 found that headstarted tortoises suffer from poor health if they lack access to certain resources like native forbs (as opposed to introduced grasses) and sufficient space for individual burrows.

Successful reintroduction or translocation of Mojave Desert Tortoises depends on judicious selection of where and how they are released. Ensuring they do not disperse too far from the release site (into unsuitable conditions) requires things like temporary holding pens, supplemental food upon release, habitat with large rocks and dry streambeds, and small mammal burrows for refuge. In adult translocations, other factors come into play. There is evidence that male tortoises introduced to an area will be rejected for mating relative to resident males, thus experiencing poor reproductive success. Post-release monitoring is essential for all release scenarios.

Despite ongoing conservation measures, the most recent five-year review of the Mojave Desert Tortoise shows continued declines from introduced predators, off-road vehicles, invasive plants, cattle, and other ongoing impacts (Chap. 9). Even with the collaboration of multiple public and private entities, shifting the extinction trajectory of tortoise populations is challenging, simply given the pervasiveness of human-wrought changes to their ecosystems. Experts recommend a multipronged conservation approach going forward, with an emphasis on restoring habitat quality to include a diverse array of native plants such as legumes where there are now invasive grasses. Because the impacts on tortoises are complex, the solutions to stemming their demise are also proving to be complex.

Looking beyond individual Mojave Desert Tortoise populations, conservation efforts are considering how to maintain longer-range movements to ensure gene flow between populations. Habitat of this species spans the Mojave and Colorado Deserts, but many populations are isolated by other land uses such as solar arrays and ranching. Since tortoises tend to traverse areas with little slope and higher vegetative cover, studies of how to maintain connectivity of their habitats focus on flatlands, valleys, and mountain passes. Certain manmade features—including railroads, flood control berms, and highways—are nearly insurmountable for tortoises. Thus ensuring tortoises population linkages will require a network of connections across suitable habitats between Tortoise Conservation Areas.

Bolson Tortoise

The related Bolson Tortoise has a much more restricted range than the Mojave Desert Tortoise. Until recently, they inhabited a small part of north-central Mexico. In 1975, the Mexican Institute of Ecology (INECOL) created a protected habitat known as the Mapimí Biosphere Reserve (RBM). Although they were still threatened by agriculture, mining, ranching, and collection for food, the RBM secured protection for about 20% of this species' range. Additional land acquisitions since have expanded the protected area. For example, about 43,000 acres were purchased by US and Mexican partners: the Rainforest Trust, HABIO A.C., and the Turtle Conservancy. The Bolson Tortoise Ecosystem Reserve, encompassing 96 mi^2, also benefits from collective management by US and Mexican conservation organizations.

Bolson Tortoises once occupied much of the Chihuahuan Desert that extends from Mexico northward into parts of Arizona, New Mexico, and Texas before harvest by pre-Colombian people in the Late Pleistocene extirpated it from those areas. Thus the reintroduction of Bolson Tortoise to the United States constitutes recovery of prehistoric populations. In 2021, the first 55 juvenile Bolson Tortoises were released in New Mexico from a captive population (a collaboration of the Turner

Endangered Species Fund, the USFWS, and New Mexico State University). Additional releases and monitoring will determine the success of the recovery effort and could set a precedent for recovery of other tortoise species to prehistoric ranges.

Whether "rewilding" habitats with tortoises extirpated in historic or prehistoric times is successful, the changing climate will come into play in shaping population outcomes. Most tortoise introductions go hand and hand with native plantings. How plants respond to climate change will determine the quality of habitat and forage available to tortoises and thus must be considered in planning for rewilding efforts. Other considerations will include availability of water, shade, and suitable temperatures and terrain.

Lessons Learned about Tortoise Conservation

Translocation

As tortoise habitats degrade under the pressures from human activities, relocations ("translocation") to national parks or other protected areas have become increasingly justified. For example, Mojave Desert Tortoises need relocation from areas slated to become active military training grounds, such as western training areas. Tortoises marked and bearing radio transmitters—as well as incidental, unmarked tortoises—are located and moved where funding permits. Capturing tortoises for translocation also gives biologists a chance to check their health and hydrate them if they've been living in drought conditions, which are frequent in desert environments (Fig. 10.4).

But relocations must take into account the genetics of the individuals to be translocated versus those of the receiving population, how to control diseases, and other aspects of successful integration. There's emerging evidence that individual heterozygosity (amount of genetic diversity in a single tortoise) plays a role in how well a tortoise survives a translocation. Tortoises with higher genetic diversity may have higher fitness for surviving in the new environments they encounter and should perhaps be prioritized in selecting which individual tortoises to translocate.

In deciding whether to introduce tortoises to novel areas, their ecological roles should be considered with an understanding that in some cases tortoises could prove effective ecological substitutes for extinct taxon. Tortoises, as intensive plant grazers and browsers, may help restore function to areas that have degraded owing to the extinction of native grazers. Even in settings like humid forest ecosystems, tortoises—as fruit eaters and seed dispersers—should be considered candidates for ecological replacement of extirpated species.

Figure 10.4
Wildlife biologist Natalie Cibel holding a Mojave Desert Tortoise (*Gopherus agassizii*) outfitted with a radio transmitter as part of the Western Ecological Research Center's Desert Tortoise research program.
Photograph courtesy of Ashley Alardi

Captive Breeding

For imperiled tortoise species, captive breeding has proved invaluable and, in some cases, the only lifeboat between them and extinction. Captive populations also provide opportunities for research on a species that can further bolster its conservation. Tortoise health in captivity requires attention to various parameters, including temperature, humidity, forage, and crowding. For example, if the density of tortoises in an enclosure gets too high, particularly of individuals mixed from different geographic locations, tortoise survival suffers. The stress of high-density captivity may cause a wasting phenomenon in which individual tortoises fail to garner sufficient nutrition. Depending on the species, males and females must be kept separate except when mating. Decades of experience with tortoises in captivity have demarcated best management practices for keeping and breeding tortoises.

In general, captive breeding programs are more successful when situated within the natural range of the tortoise species (Plate 24), simply because the conditions of microclimate, seasonality, and day length are a priori suitable for the target species. The captive animals are likely to already have disease resistance to local pathogens and, in the case of reintroductions, are less likely to transmit diseases to wild populations. In some cases—for species that are easily kept in captivity—a reservoir of tortoises exists as pets for people who might cooperate with breeding programs, thus garnering more genetic diversity. Zoos and veterinary facilities outside of tortoises' native range can provide support that may otherwise not be available in-country to breeding programs.

Headstarting

Although tortoises have high adult survival, hatchlings and juveniles are vulnerable to a range of predators (Chap. 5). The focus of conservation in the wild has been on adult tortoises because of their reproductive potential (waiting for hatchlings to reach reproductive age takes years). Headstarting turtles has become a viable approach to ensure a continued flow ("recruitment") of new individuals into a population.

Hatchlings may come from wild-collected eggs or from eggs laid by captive adults. In either case, because most, if not all, tortoises have temperature-dependent sex determination, headstarting strategies must ensure suitable ratios of males and females (Chap. 4). Because sex characteristics may not be apparent in hatchlings for several years until they approach maturity, research data must be used to determine appropriate incubation temperatures. The pivotal temperature for males versus females will depend on the temperature regime of the tortoise species native range.

In headstarting, there's a trade-off between time spent raising an individual tortoise and its susceptibility to predation. These factors are negatively correlated because larger tortoises survive better (Chap. 3). The costs of rearing tortoises and the condition of the habitat they are released into—especially regarding introduced predators—determine the optimal length of any headstarting program. Methods to accelerate growth in captivity, such as higher temperatures and more forage, can reduce the requisite headstarting time or the risks of predation upon release. A rough rubric used by headstarting programs is that juvenile tortoises should reach at least 10 cm carapace length before release because beyond that size they become more difficult for a predator to eat.

Tortoises are innately programmed to fend for themselves from birth, with no parental care, so they don't need to be trained before release. But raising multiple generations in captivity could lead to a loss of the genetic adaptation that allows them to survive in their natural habitats. To maintain innate capabilities in relation to their environments, breeding and release programs should avoid holding on to multiple generations of a species.

Habitat Protection

As the habitats that tortoises depend on continue to be altered for various human uses, imperiled tortoise species need portions of habitat of suitable sizes to sustain populations. Habitat fragmentation reduces the quality of habitat, encourages invasive species, and affects genetic diversity. Research that maps the distribution and abundance of species across their entire range can provide the needed data to protect critical habitat blocks. Given the density and intensity of human activities, it's no longer possible to protect the entire native range of any tortoise species. Instead, efforts must be directed toward establishing reserves that protect some of a species' best populations and ensure that habitat conditions remain suitable for their long-term survival.

Moreover, data on genetic variation and population structure of tortoise species provide a necessary conservation input. Looking at how tortoise population genetics vary across the landscape reveals interactions between the environment and evolutionary processes such as genetic change through natural selection. The objective of a well-informed conservation program is to maintain genetic diversity through natural connectivity between protected populations as well as engineered variation via captive rearing and release. Failure to foster sufficient genetic diversity in tortoises leaves populations vulnerable to introduced pathogens or predators. For example, disease can ripple across all remaining populations. Genetic diversity confers resilience to change, and the field of "landscape genomics" helps ensure it.

Local Engagement

Efforts to protect tortoise habitat are only as effective as their implementation and enforcement. Local people who use the resources around them are the first-order consumers of tortoises and must be on board with their protection. The number of tortoises consumed locally may pale compared to regional and international export numbers. Still, all consumptive uses reduce the viability of populations over time, and local people may capture and sell tortoises to wildlife brokers (Chap. 9). Yet in Serbia, where Hermann's Tortoises are collected for foods, medicines, and pets, barely over half of people living within the boundaries of national parks report knowing they reside in protected areas. Linking conservation actions to alleviation of the factors driving exploitation, while cultivating a well-informed populace, opens avenues for sustained protection of tortoises.

Furthermore, traditional values regarding tortoises can be brought to bear on conservation planning. The Antandroy and Mahafaly Peoples of Madagascar, for example, maintain taboos on consumption of tortoises that continue to benefit

Radiated Tortoise populations and could be bolstered through broader education and support. The taboos are grounded in ecological knowledge of tortoises that's inherent to living among them. One study respondent in the Androy region of Madagascar said, "A hand has 5 fingers and when 2 fingers are missing, it is not a well-functioning hand anymore. So, when the tortoises are gone, nothing will work anymore." This appreciation for the keystone role of tortoises renders local people powerful actors in their conservation. As people migrate into these communities from urban areas or other places that lack direct experience with tortoises, local leaders must educate, incentivize, and otherwise pave the way for tortoise protection.

Long-Term Studies

For populations of animals with shorter life spans, like insects or small mammals, studies over several months or years can provide sufficient data for management. For tortoises, however, their slow development, late reproduction, and prolonged life spans make long-term studies essential (Chap. 5). Studies that span decades provide a view of population dynamics and adaptations to fluctuating habitat conditions. They also encompass lag times when tortoise populations (because of their drawn-out life histories) are slow to respond to change; that is, a positive or negative change in conditions may take a long time to show up in the demography of the population. A population of aging adults may survive for years after successful reproduction is no longer occurring.

As basic life history data (such as age at first reproduction and survival rates of distinct age classes) accumulate for tortoises, predictive modeling becomes a feasible tool for conservation planning. Habitat models, assuming sufficient data, can reliably predict occurrences of tortoise populations based on habitat resources, and thus predict responses to changes in those resources. It will be impossible to stop the whole spate of ongoing threats to every species of tortoise. Predictive data on how each threat affects populations can help prioritize which threats to reduce or eliminate.

In the face of climate change, modeling how tortoise populations are likely to respond is key to developing conservation plans. Extreme weather is expected to become increasingly common and unpredictable. Although tortoises are sufficiently ancient to have evolved a range of strategies to cope with environmental extremes (Chap. 3), they do have limits to their adaptability. One of the main climate threats to tortoises will be the increased frequency of severe droughts. Coupled with high temperatures that may affect sex ratios, droughts could extirpate tortoise populations unless their impacts can be somehow mitigated.

Collaboration

The overarching lesson learned to date is that it will take collaborations of people and organizations to keep imperiled tortoise species on Earth. The complex interactions of threats to tortoises call for an equally complex strategy to address them. In every successful ongoing conservation program, there are a suite of roles shared across public and private organizations, with motivated individuals in the lead, including people living within the native ranges of tortoise species.

Despite their longevity on Earth and demonstrated resilience through dramatic changes over millions of years, tortoises will need all the advocates they can garner to remain on this planet during the unprecedented pressures they face in the Anthropocene Age. With enough knowledge and commitment from humans, some tortoise species will survive, the adaptable progeny of an ancient lineage of reptilian survivors.

ACKNOWLEDGMENTS

No book can be written without the assistance and encouragement of colleagues. We are especially grateful and appreciative of that assistance from the individuals mentioned below and the publications (*Herpetologica*, *Journal of Herpetology*, and *Western North American Naturalists*) that gave us permission to use illustrations and photographs. We also wish to thank Google Scholar, ResearchGate, and the Smithsonian Archives and Library for access to the scientific literature on tortoises.

We thank Stephen Platt and the *G. platynota* restoration team for size information on their adult breeding. A personal thank-you to Kenneth Dodd for his explanation of the US endangered species legislation. David Addison, Harry Greene, Thomas Reese, and Ian Swingland graciously read all the chapter manuscripts, ferreted out errors, and otherwise improved the text. A special shout-out of thanks to them. A big thank-you to Evan Vlachos for bringing our fossil chapter and its nomenclature up to date (as of September 2022) and to Thomas Leuteritz and Aryeh Miller for reading select chapters. Thanks to all!

The following colleagues gave us permission to use their tortoise photographs in this book: Natalie Cibel, Kenneth Dodd, Cris Hagen, Catharine Joynson-Hicks, Thomas Leuteritz, J. C. Schaffer, and Ian Swingland. Other photographs either in the public domain or licensed to the Creative Commons derive from online sources, especially Wikipedia Commons. We appreciate the generosity of all for allowing us to highlight their photographs.

All line illustrations are by Catalina Montalvo unless otherwise noted. Preferring the clarity of line illustrations over photographic ones to highlight specifics of anatomy and behavior, we asked Ms. Montalvo to prepare illustrations, mostly from photographs and sketches provided by George. This collaboration was a long-distance one between her studio in southwestern Florida and our metro Washington, DC, offices. We also extracted images from publications to be redrawn and/or reoriented, noting the literature sources and obtaining permission for those figures still under copyright.

FURTHER READINGS

Bonin, F., B. Devaux, and A. Dupré. 2006. *Turtles of the World.* Johns Hopkins University Press.

Branch, B. 2008. *Tortoises, Terrapins and Turtles of Africa.* Penguin Random House South Africa.

Chambers, P. 2005. *A Sheltered Life: The Unexpected History of the Giant Tortoise.* Oxford University Press.

Ernst, C., and J. Lovich, 2009. *Turtles of the United States and Canada.* Johns Hopkins University Press.

Gibbs, J., et al. 2020. *Galapagos Giant Tortoises.* Academic Press.

Hennessy, E. 2019. *On the Backs of Tortoises: Darwin, the Galapagos, and the Fate of an Evolutionary Eden.* Yale University Press.

Holger, V. 2005. *TERRALOG: Turtles of the World.* Vols. 1–4. Edition Chimaira.

Lovich, J. E., and W. Gibbons. 2021. *Turtles of the World: A Guide to Every Family.* Princeton University Press.

Orenstein, R. 2012. *Turtles, Tortoise and Terrapins: A Natural History.* Firefly Books.

Rhodin, A. G. J., J. B. Iverson, R. Bour, U. Fritz, A. Georges, H. B. Shaffer, and P. P. van Dijk. 2021. *Turtles of the World: Annotated Checklist and Atlas of Taxonomy, Synonymy, Distribution, and Conservation Status.* 9th ed. Chelonian Research Foundation and Turtle Conservancy.

Rose, F. L., and F. W. Judd. 2014. *The Texas Tortoise: A Natural History.* University of Oklahoma Press.

Stanford, C. B. 2010. *The Last Tortoise: A Tale of Extinction in Our Lifetime.* Belknap Press of Harvard University Press.

The full research bibliography for this book is available within the Smithsonian Herpetological Information Services' website: https://repository.si.edu/handle/10088/842.

APPENDIX

Turtle Families and Tortoise Species

Table A.1.
Classification of Living Turtles (Order Testudines)

Classification	Standard English Name
Suborder Cryptodira	**Hidden-neck Turtles**
Carettochelyidae	Pig-nosed Turtles
Cheloniidae	Hard-shelled Sea Turtles
Chelydridae	Snapping Turtles
Dermatemydidae	Central American River Turtles
Dermochelyidae	Leatherback Sea Turtles
Emydidae	Emydid Turtles
Kinosternidae	Mud and Musk Turtles
Geoemydidae	Geoemydid Turtles
Platysternidae	Big-headed Turtles
Testudinidae	Tortoises
Trionychidae	Softshell Turtles
Suborder Pleurodira	**Side-neck Turtles**
Chelidae	Australian–South American Side-neck Turtles
Pelomedusidae	African Side-neck Turtles
Podocnemidae	Neotropical Side-neck River Turtles

Table A.2.
Tortoise Taxa

Scientific Names	Standard English Names	IUCN Redbook Status as of September 30, 2022
Aldabrachelys abrupta	Madagascar Giant Tortoise	Extinct
Aldabrachelys gigantea	Aldabra Giant Tortoise	Vulnerable
Aldabrachelys grandidieri	Grandidier's Giant Tortoise	Extinct
Astrochelys radiata	Radiated Tortoise	Critically Endangered
Astrochelys rogerbouri	Bour's Tortoise	Extinct
Astrochelys yniphora	Ploughshare Tortoise	Critically Endangered

Table A.2.
Tortoise Taxa (*cont.*)

Scientific Names	Standard English Names	IUCN Redbook Status as of September 30, 2022
Centrochelys sulcata	African Spurred Tortoise	Endangered
Chelonoidis carbonarius	Red-footed Tortoise	Not Evaluated
Chelonoidis chilensis	Chaco Tortoise	Vulnerable
Chelonoidis denticulatus	Yellow-footed Tortoise	Vulnerable
Chelonoidis niger niger	Floreana Giant Tortoise	Extinct
Chelonoidis n. abingdonii	Pinta Giant Tortoise	Extinct
Chelonoidis n. becki	Volcán Wolf Giant Tortoise	Vulnerable
Chelonoidis n. chathamensis	San Cristóbal Giant Tortoise	Endangered
Chelonoidis n. darwini	Santiago Giant Tortoise	Critically Endangered
Chelonoidis n. donfaustoi	Cerro Fatal Giant Tortoise	Critically Endangered
Chelonoidis n. duncanensis	Pinzón Giant Tortoise	Critically Endangered
Chelonoidis n. guntheri	Sierra Negra Giant Tortoise	Critically Endangered
Chelonoidis n. hoodensis	Española Giant Tortoise	Critically Endangered
Chelonoidis n. microphyes	Volcán Darwin Giant Tortoise	Endangered
Chelonoidis n. phantasticus	Fernandina Giant Tortoise	Critically Endangered
Chelonoidis n. porteri	Western Santa Cruz Giant Tortoise	Critically Endangered
Chelonoidis n. vandenburghi	Volcán Alcedo Giant Tortoise	Vulnerable
Chelonoidis n. vicina	Cerro Azul Giant Tortoise	Endangered
Chersina angulata	Angulate Tortoise	Least Concern
Chersobius boulengeri	Karoo Dwarf Tortoise	Endangered
Chersobius signatus	Speckled Dwarf Tortoise	Endangered
Chersobius solus	Nama Dwarf Tortoise	Vulnerable
Cylindraspis indica	Reunion Giant Tortoise	Extinct
Cylindraspis inepta	Mauritius Giant Domed Tortoise	Extinct
Cylindraspis peltastes	Rodrigues Domed Tortoise	Extinct
Cylindraspis triserrata	Mauritius Giant Flat-shelled Tortoise	Extinct
Cylindraspis vosmaeri	Rodrigues Giant Saddleback Tortoise	Extinct
Geochelone elegans	Indian Star Tortoise	Vulnerable
Geochelone platynota	Burmese Star Tortoise	Critically Endangered
Gopherus agassizii	Mojave Desert Tortoise	Critically Endangered
Gopherus berlandieri	Berlandier's Tortoise	Least Concern

Scientific Names	Standard English Names	IUCN Redbook Status as of September 30, 2022
Gopherus evgoodei	Sinaloan Thornscrub Tortoise	Vulnerable
Gopherus flavomarginatus	Bolson Tortoise	Critically Endangered
Gopherus morafkai	Sonoran Desert Tortoise	Not Evaluated
Gopherus polyphemus	Gopher Tortoise	Vulnerable
Homopus areolatus	Parrot-beaked Dwarf Tortoise	Least Concern
Homopus femoralis	Greater Dwarf Tortoise	Least Concern
Indotestudo elongata	Yellow-headed Tortoise	Critically Endangered
Indotestudo forestenii	Sulawesi Tortoise	Critically Endangered
Indotestudo travancorica	Travancore Tortoise	Vulnerable
Kinixys belliana	Bell's Hinge-back Tortoise	Not Evaluated
Kinixys erosa	Forest Hinge-back Tortoise	Data Deficient
Kinixys homeana	Home's Hinge-back Tortoise	Critically Endangered
Kinixys lobatsiana	Lobatse Hinge-back Tortoise	Vulnerable
Kinixys natalensis	Natal Hinge-back Tortoise	Vulnerable
Kinixys nogueyi	Western Hinge-back Tortoise	Not Evaluated
Kinixys spekii	Speke's Hinge-back Tortoise	Not Evaluated
Kinixys zombensis	Southeastern Hinge-back Tortoise	Not Evaluated
Malacochersus tornieri	Pancake Tortoise	Critically Endangered
Manouria emys	Asian Giant Tortoise	Critically Endangered
Manouria impressa	Impressed Tortoise	Endangered
Psammobates geometricus	Geometric Tortoise	Critically Endangered
Psammobates oculifer	Serrated Tent Tortoise	Not Evaluated
Psammobates tentorius	Tent Tortoise	Not Threatened
Pyxis arachnoides	Spider Tortoise	Critically Endangered
Pyxis planicauda	Flat-tailed Tortoise	Critically Endangered
Stigmochelys pardalis	Leopard Tortoise	Least Concern
Testudo graeca	Spur-thighed Tortoise	Vulnerable
Testudo hermanni	Hermann's Tortoise	Not Threatened
Testudo horsfieldii	Steppe Tortoise	Vulnerable
Testudo kleinmanni	Egyptian Tortoise	Critically Endangered
Testudo marginata	Marginated Tortoise	Least Concern
47 extant species		

STANDARD AND SCIENTIFIC NAME INDEX

Most tortoises have several standard English names. We use only a single name for each species. For consistency and standardization with the current scientific and conservation literature on tortoises, we usually selected the first standard English name given in the Conservation Biology of Freshwater Turtles and Tortoise 2021 Checklist, with a few exceptions.

Species marked with an asterisk (*) are recently extinct (historical time).

SUBJECT INDEX